Principles of
Chemical Sensors

MODERN ANALYTICAL CHEMISTRY

Series Editor: David Hercules
University of Pittsburgh

ANALYTICAL ATOMIC SPECTROSCOPY
William G. Schrenk

APPLIED ATOMIC SPECTROSCOPY
Volumes 1 and 2
Edited by E. L. Grove

CHEMICAL DERIVATIZATION IN ANALYTICAL CHEMISTRY
Edited by R. W. Frei and J. F. Lawrence
Volume 1: Chromatography
Volume 2: Separation and Continuous Flow Techniques

COMPUTER-ENHANCED ANALYTICAL SPECTROSCOPY
Volume 1: Edited by Henk L. C. Meuzelaar and Thomas L. Isenhour
Volume 2: Edited by Henk L. C. Meuzelaar

ION CHROMATOGRAPHY
Hamish Small

ION-SELECTIVE ELECTRODES IN ANALYTICAL CHEMISTRY
Volumes 1 and 2
Edited by Henry Freiser

LIQUID CHROMATOGRAPHY/MASS SPECTROMETRY
Techniques and Applications
Alfred Yergey, Charles Edmonds, Ivor Lewis, and Marvin Vestal

MODERN FLUORESCENCE SPECTROSCOPY
Volumes 1–4
Edited by E. L. Wehry

PHOTOELECTRON AND AUGER SPECTROSCOPY
Thomas A. Carlson

PRINCIPLES OF CHEMICAL SENSORS
Jiří Janata

TRANSFORM TECHNIQUES IN CHEMISTRY
Edited by Peter R. Griffiths

Principles of
Chemical Sensors

Jiří Janata

University of Utah
Salt Lake City, Utah

PLENUM PRESS • NEW YORK AND LONDON

Library of Congress Cataloging in Publication Data

Janata, Jiří.
    Principles of chemical sensors / Jiří Janata.
        p.   cm. — (Modern analytical chemistry)
    Includes bibliographical references and index.
    ISBN 0-306-43183-1
    1. Chemical detectors. I. Title. II. Series.
TP159.C46J36   1989   89-8695
681'.2 — dc20   CIP

TP
159
.C46
J36
1989

© 1989 Plenum Press, New York
A Division of Plenum Publishing Corporation
233 Spring Street, New York, N.Y. 10013

Printed in the United States of America

To the memory of my teachers

Preface

It is sometimes difficult to appreciate the impact of events taking place
in the period of time within which we live and work. Perhaps one day
people will look back in awe at the second half of this century and marvel
at the extent of the computer revolution which took place at that time.
Whether we like it or not, we are all very much a part of it, regardless of
our specialization. The sheer amount of data-processing capacity which
exists now and is growing out of all proportion creates a need for new
information from every possible, and sometimes impossible, source. The
purpose of a sensor is to provide information about our physical, chemical,
and biological environment. In that respect it is a logical element in the
information acquisition–processing chain.

The field of chemical sensors is one of the most diverse I have ever
encountered, owing to the nature of the environment sensed and the
different types of processes involved. This creates obvious problems in
writing a textbook on the subject. If one invites coauthors the final result
should be profound in every aspect covered, but it is rarely presented
evenly. In preparing a single-author textbook, the writer inevitably covers
subjects which lie outside his area of expertise and can only provide his
own interpretations. I hope that the resulting number of serious errors will
be minimal. The second approach has been selected mainly for two reasons.
In the last few years I have been teaching courses on chemical sensors at
various levels. To convert lecture notes to a textbook is seemingly natural,
particularly in the absence of other suitable texts in the area. The second
reason is more fundamental. It is possible to engineer a new sensor in such
a way that its output is related to the concentration for a given set of
conditions without paying too much attention to the mechanism of the
processes involved. The result is a device which may perform well within the
conditions for which it has been developed, but *only* for those conditions.

The danger is that the result so obtained may be an experimental artifact. Another approach is to examine in as much detail as possible the principles underlying the operation of a new device. This may not lead to a new sensor immediately, but those developed along these lines tend to be more reliable. The accent in this book is therefore on the principles behind the operation ("the trade") rather than on a description of applications ("the tricks of the trade") of individual sensors. In this respect it is written for students at both graduate and upper undergraduate levels. Approximately one semester's worth of material is presented. The book may also be useful for scientists and engineers involved in the development of new types of chemical sensors or for those who discover that "somebody else's sensor just does not work as it should" and wish to know why.

The book is divided into five sections dealing with the four principal modes of transduction: thermal, mass, electrochemical, and optical, as well as a general introduction common to the four types. I have included five appendixes, which are intended as a quick reference for readers who may not possess sufficient background in some areas covered in the main text. I have run out of symbols in both the Latin and Greek alphabets. In order to avoid confusion and ambiguity I have confined the use of a set of symbols to each chapter and provided glossaries at the end of each chapter.

Finally, I wish to thank all my students and collaborators at the University of Utah from whom I have learned this trade, and also, in particular, my secretary of many years, Mrs. LaRue Dignan. I furthermore thank my colleagues and friends at the Universität der Bundeswehr München who provided the environment in which most of this book was written.

Jiří Janata

Salt Lake City and Munich

Contents

Appendixes 285

Principles of
Chemical Sensors

General Aspects

1.1. Introduction

Chemical sensing is part of an information acquisition process in which some insight is obtained about the chemical composition of the system in real time. In this process an amplified electric signal results from the presence of some chemical species. Generally, it consists of two distinct steps: *recognition* and *amplification.* An example is the ordinary measurement of pH with a glass electrode (Figure 1-1). The interaction of the hydronium ion with the electrode is highly specific, but the power in the primary electric signal is very low. For a 10 MΩ glass electrode and 1 mV error it would be approximately 1 pW. If we try to draw more power from such an electrode, the information would be distorted or destroyed. In other words, the source of the signal (electrode) requires an amplifier (pH meter) in order to obtain the information in a useful, undistorted form. Thus the recognition (selectivity) is provided by some chemical interaction while the amplification can be provided by some physical means. There are exceptions, however, to this statement: for example, enzymatic reactions combine the high selectivity of the enzyme binding for a given substrate with catalytic properties of the enzyme that represent an amplification step in its own right.

The coupling of the chemically selective layer to the physical part of the sensor is very important. As we shall see it can have a profound effect on chemical selectivity. The signal can subsequently be manipulated in many different ways and with different degrees of sophistication. Thus it can be displayed in analog form, subtracted from the reference signal and displayed as a difference, or it can be digitized or processed statistically, etc. Such processing can be carried out within the physical boundary of the

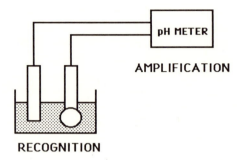

AMPLIFICATION

RECOGNITION

Figure 1-1. Recognition and amplification in chemical sensing.

sensor itself or in a separate processor. In the former case it is fashionable to talk about "smart sensors."

With so many different sensors described in the literature one can ask: which one is the best? There are so many different sensing situations and criteria to be considered that the general answer is: the sensor which will do the job. There is, however, one characteristic of a chemical sensor which sets it apart from the others and that is its "ruggedness." This term describes the ability of the device to maintain its performance specifications even under adverse operating conditions. In a pragmatic sense it is understood as "reliability."

Ruggedness may have different meanings; a mechanically rugged device is able to withstand mechanical shock, vibrations, mechanical stresses, and so on. Chemical ruggedness has a slightly more subtle meaning. It is related to selectivity and means that the output of the sensor is unaffected by unforeseen chemical changes in the operating environment. Generally speaking, rugged sensors are those commercially most successful because they are reliable or because they are the only ones which can provide information under conditions in which other sensors would be unable to operate. One example could be a potentiometric sensor for measuring oxygen concentration in molten steel. There are not many other sensors which could do that particular job. It is usually useful to assess a "new sensor" on the basis of its "ruggedness" as required by the given operating conditions.

It is generally possible to distinguish two types of interaction of the chemical species with the sensor: a *surface interaction* in which the species of interest is *adsorbed* at the surface, and a *bulk interaction* in which the species of interest is *absorbed* and *partitions* between the sample phase and the sensor. The decrease of the free energy is the driving force in all sensing processes.

Thermodynamical reversibility does not have any special meaning in the classification of individual sensors. Thus, sensors can be either thermodynamically reversible (such as ion-selective electrodes) or nonreversible (such as enzyme electrodes or amperometric sensors), in which a *steady-state* response is obtained to a step up or down in the concentration of the substrate. Likewise, detectors can be reversible or irreversible, but dosimeters are almost always irreversible. However, thermodynamical considerations are important in understanding the principles governing the operation of the individual devices.

The interaction of a chemical species X with the sensor S can be described by the equilibrium

$$X + S \underset{k_r}{\overset{k_f}{\rightleftharpoons}} XS \tag{1-1}$$

The equilibrium constant K expressed in terms of *activities*

$$a = fC \tag{1-2}$$

is given by the equation

$$K = a_{sx}/a_s a_x = k_f/k_r \tag{1-3}$$

where a_{sx} is the activity of the bound species, while a_x and a_s are the activities of the species in the sample and of the binding site, respectively. The overall rates of the forward (k_f) and reverse (k_r) reactions together with the mass-transport parameters of the species involved in the trans-duction mechanism determine the time response of the sensor.

The free energy of interaction for the reaction shown in equation (1-1) is [see equations (A-17) and (A-18) in Appendix A at the end of this book]

$$\partial G = 0 = \partial G^\circ + RT \ln(a_{sx}/a_s a_x) \tag{1-4}$$

It follows from equation (1-4) that, upon a change in sample activity, interaction of the species with the sensor will take place if the variation in the standard free energy ∂G° is negative and/or the argument in the logarithmic term is less than unity. If the binding equilibrium constant is too high $(K > 10^4; \partial G^\circ < 5.5 \text{ kcal mol}^{-1})$ the reaction will be nearly irreversible from a practical point of view, and the device will respond in a nonequilibrium manner, as a dosimeter.

From the sensor application standpoint, the high value of the binding constant is desirable but the virtually irreversible nature of the interaction is a problem. This may seem paradoxical because, intuitively, one would feel that a high value of the binding constant is desirable. It certainly is in the case of, for example, ion-selective electrode membranes, so why should this be a problem? Let us assume that only the bound species SX produces the output signal. The total activity of binding sites in the sensor is given by

$$a_T = a_{SX} + a_S \qquad (1-5)$$

Equations (1-3) and (1-5) yield

$$a_{SX} = K a_T a_X / (1 + K a_X) \qquad (1-6)$$

For $K a_X \gg 1$, all the available sites become occupied and $(a_{SX})_{max} = a_T$. Hence the sensor reaches its *saturation*, i.e., the upper limit of its dynamic range. On the other hand, for $K a_X \ll 1$,

$$a_{SX} = K a_T a_X \qquad (1-7)$$

and the sensor operates within its dynamic range. There is a lower limit of the activity of bound species, $(a_{SX})_{min}$, below which the sensor does not respond. Its absolute value obviously depends on the physics of the transduction mechanism and is usually set as that value of the signal equal to its two standard deviations. The dynamic range is therefore given by the difference between these two values,

$$\text{dynamic range} = (a_{SX})_{max} - (a_{SX})_{min} \qquad (1-8)$$

For a given type of sensor the minimum value is usually invariable, which means that the dynamic range depends on the total number of binding sites a_T. For example, a sensor with a detectable minimum density of 10^9 molecules cm^{-2} and a maximum density of 10^{14} sites cm^{-2} (corresponding to 100 Å^2 molecule^{-1}) has a theoretical dynamic range of five decades. How is it then possible that, for example, a glass electrode has a dynamic range for hydrogen ion activity extending over *thirty* decades? There are at least two reasons. First, the number of binding sites is much larger than 10^{14} sites cm^{-2}. This can happen if, instead of being in a single plane, the binding sites are distributed throughout a layer of finite thickness. This directly increases a_T, and thus $(a_{SX})_{max}$, by two orders of magnitude. The second reason, and more important factor, is the existence of different multiple binding constants $K_1, K_2, ..., K_n$ for the species X. In that case the response will include a contribution from all the binding constants.

Let us assume that the output E_{out} of the sensor is related to a_{SX} by a general response function $\Re$, i.e.,

$$E_{out} = \Re(a_{SX}) \qquad (1-9)$$

According to equation (1-7) the activity of the bound species is given by

$$a_{SX} = a_X(K_1 a_{T,1} + K_2 a_{T,2} + \cdots + K_n a_{T,n}) \qquad (1\text{-}10)$$

where $a_{T,1}, a_{T,2},...$ are the total activities of the individual binding sites. If the partial binding processes do not affect each other, the overall response has the functional relationship $\Re$ which is characteristic of the individual binding constant,

$$E_{out} = \Re a_X(K_1 a_{T,1} + K_2 a_{T,2} + \cdots + K_n a_{T,n}) \qquad (1\text{-}11)$$

Such a situation exists within the *gel layer* of a glass electrode with multiplicity of binding sites for hydrogen ion provided by the statistically large number of lattice defects and doping impurities. They yield a series of overlapping dynamic ranges with the overall span much larger than would be achieved by any single binding site with only one equilibrium constant. In that case the function $\Re$ is $RT/zF \ln(\quad)$. We can see that a *surface* with a *single* binding constant has only limited theoretical dynamic range. This is true of any sensor whose primary transduction mechanism relies only on surface interactions.

The above discussion has been formulated in terms of the free energy but did not consider the *types* of chemical interaction which may be involved. They are summarized in Table 1-1, which shows the energies of interactions that determine the selectivity.[1]

The location of the sites is important from the viewpoint of the

Table 1-1. Approximate Interaction Enthalpies[a]

Type of interaction	Equation	Order of magnitude (kcal mol^{-1})
Covalent bond	—	50–200
Hydrogen bond	—	1–10
Ion–ion	$E \sim z_1 z_2/Dr$	10–100
London forces	$E \sim \alpha_1 \alpha_2/r^6$	1–10
Ion-induced dipole	$E \sim z_1^2 \alpha_2/Dr^4$	0.1–1
Dipole–dipole	$E \sim \mu_1 \mu_2/Dr^3 kT$	0–1
Hydrophobic bond	$CH_3 CH_3$	(0.3)*
		(1.5)*

[a] D is dielectric constant; α is polarizability; μ is dipole moment; r is distance; z is charge; * means per mole unit coupling.

operational characteristics. Sensors whose response depends on *bulk inter-actions* are based on the Gibbs equation (A-20) which, for a two-phase equilibrium, is

$$\partial G = 0 = (\mu_i \, dn_i)_{\text{sensor}} + (\mu_i \, dn_i)_{\text{sample}} \tag{1-12}$$

where i is the chemical species which partitions between the sample and sensor phase. At equilibrium the number of moles of i crossing the interface must be equal (but of opposite sign):

$$(dn_i)_{\text{sensor}} = -(dn_i)_{\text{sample}} \tag{1-13}$$

Thus the chemical potentials of species i in the two phases must be equal,

$$(\mu_i)_{\text{sensor}} = (\mu_i)_{\text{sample}} \tag{1-14}$$

The definition of the chemical potential in a real phase (gas, solid, or solution) is [cf. equations (A-16), (A-19), and (A-24)]

$$\mu_i = \mu_i^0 + RT \ln a_i \tag{1-15}$$

Thus, sensors based on absorption (phase equilibrium) measure *activity* and their output is (usually) *logarithmic*.

The condition of the general adsorption equilibrium which does not assume the existence of high-affinity binding sites is again described by the equality of the chemical potential of the species in the sample phase (μ_X) and at the surface (μ_{SX}). If the adsorbing species carry a formal electric charge, then the equilibrium is expressed by the respective electrochemical potentials. The general adsorption equilibrium has the form of equation (1-1), and the equilibrium constant K can be expressed in a similar manner to the above:

$$X \Leftrightarrow SX \tag{1-16}$$

and

$$a_X K = a_{SX} \tag{1-17}$$

where SX designates the adsorbed species. We define the surface concentration as Γ^i moles per area, which is related to the surface activity through some form of adsorption isotherm. The general adsorption isotherm is given by equations (A-27) and (A-28),

$$\beta c = F(\Gamma) \tag{1-18}$$

where

$$\beta = \exp(-\varDelta G_a/RT) \qquad (1\text{-}19)$$

The free energy of adsorption $-\varDelta G_a$ depends on the type of substrate–adsorbate and adsorbate–adsorbate interactions. Again, the interaction equilibria are formulated in the form of *activities* but the response function of the sensor is dictated by the adsorption isotherm.

What is it then that we actually measure: activity or the concentration of the species? We can take a very uncompromising view of this question and assume that all chemical interactions are governed by activities which can be equated with concentrations only under special circumstances. In gases where the interactions between gas molecules are relatively weak, the activities (more appropriately, fugacities) can be equated with concentrations (partial pressures). Another "safe" approximation of activity to concentration is commonly made in aqueous solutions in which the ionic strength does not exceed 1 mM. For electrically neutral species the concentration of other solutes can be even higher. The immediate consequences of making this assumption without proper justification are the nonspecific interferences due to the variation in the activity coefficient, namely, the loss of the information content of the signal.

There are, however, sensors in which the output can be justifiably related to concentration even under conditions where activity and concentration differ substantially. These are sensors whose output is governed by the gradient or by the time variation of concentration, i.e., in steady-state sensors for which the signal S is given by

$$S = \Re\left[\sum_i K_i(dC_i/dx)\right] \quad \text{or} \quad S = \Re\left[\sum_i K_i(dC_i/dt)\right] \quad (1\text{-}20)$$

If we differentiate the activity equation $a = fC$ with respect to time or distance, then we obtain

$$\frac{da}{dx} = \frac{df}{dx}C + f\frac{dC}{dx} \quad \text{or} \quad \frac{da}{dt} = \frac{df}{dt}C + f\frac{dC}{dt} \qquad (1\text{-}21)$$

The variations in activity coefficients df/dx or df/dt are almost always close to zero and/or independent of the concentration of other solutes. Therefore the first terms in equation (1-21) can be neglected. In the case of first-order kinetic equations [equations (B-1)], which are by far the most common, the activity coefficients simply cancel out. Thus care must be taken only in the case of higher-order kinetics (see Appendix B).

The activity coefficients also cancel out in the second Fick's law if we accept that $df/dt = d^2f/dx^2$. Therefore, mass-transport sensors (e.g., amperometric sensors) can also be expressed in terms of concentrations rather than activities.

On the other hand, equilibrium sensors which depend on partitioning or adsorption provide information about activities, not concentrations. A special problem exists in *equilibrium optical* and *mass* sensors in which the transduction is always governed by counting the molecules, i.e., the *concentration*, but the chemical interaction is governed by *activities*.

1.2. Selectivity

Selectivity can be defined as the ability of a sensor to respond primarily to only one species in the presence of other species. In our discussion of man-made sensors it is helpful to keep is sight the ultimate sensing machine—the living organism.[2] Each biological species has three main missions in order to survive: to metabolize, to reproduce, and to process information. The latter means both information acquisition and transmission, and acquisition is what sensing is about. In the animal kingdom the sensory information is usually transformed into a simple action: "run," "fight," "have sex," "sell stock," etc. There are thousands of chemical species in the environment, and the *selectivity* is of primary importance because the acquisition of "false signals" and/or their wrong interpretation could be anything from humorous to disastrous.

Biological strategy is to involve *shape recognition*, in other words, stereospecificity. The energies in Table 1-1 are listed as *enthalpies H*, but the driving forces in the chemical species–sensor interactions are really the changes of free energy. At constant temperature the two are related by equation (A-13). The higher the entropy change, the more negative the free-energy change and the more stable the system. Consequently, let us look at the binding site S (receptor) and the chemical species of interest X in Figure 1-2 drawn intentionally in such a way that they fit (meaning: "are shape-recognized"). Within the closed system (a), they are clearly in a more random configuration if *they are not associated*, because they have more degrees of freedom. This means that by entropy considerations alone, they would be more stable if they were apart. The only way they would be used in an active sensor is if the negative enthalpy change due to their association were high enough to compensate for the value of the term *TS* in equation (A-13). Clearly, in that case the temperature would have a strong effect on the value of the equilibrium constant. In this case (Figure 1-2) the "shape recognition" strategy will work only if the negative enthalpy producing binding sites (exothermic) is present within the "cleft" of the binding site S. Nature has, indeed, employed this strategy in using

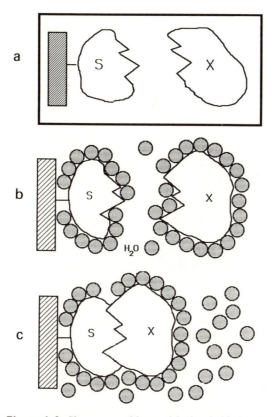

Figure 1-2. Shape recognition and hydrophobic bond.

hydrogen bonds, ionic bonds, charge transfer complexes, etc., inside the binding sites of hormone receptors, antibodies, etc. However, it can be seen from Table 1-1 that most of the interactions fall off rapidly with distance between the interacting entities (with the exception of the ion–ion bond) and with increased dielectric constant. Furthermore, the enthalpy inter-actions are common to classes of compounds and would not provide sufficient selectivity. So, is there a way out?

Let us consider the most regular object, a sphere. It has only one parameter which affects its potential to be selectively recognized, its size. There is only one "arrangement" A which corresponds to a selective fit of the sphere to the binding site. If we now take two spheres connected by a solid rod (like a dumbbell), the number of possible arrangements increases to $A = 3$ (two for the sizes of the balls and one for their separation). We take one more step in this progression (Figure 1-3) and interconnect three balls with two links of different length. For this "molecule" $A = 12$: three

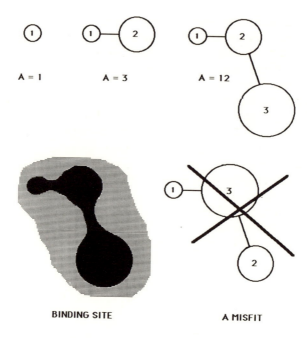

Figure 1-3. Balls, sticks, and selectivity.

for the diameters of the balls, two for the lengths of the links, one for the angle. Furthermore we can place, in turn, spheres 1, 2, and 3 in the center position. In each case the remaining two spheres can be connected by either link a or b, which gives a total of six different combinations. It is important to realize that only one configuration will fit the binding site. Generally, as the geometrical complexity increases, the number of misfits also increases. Furthermore, the "size of the ball" can be also interpreted as the "type of interaction." Therefore, a molecule approaching the binding site can be either attracted ($\Delta G < 0$) or repulsed ($\Delta G > 0$). Thus, even if the molecule fits the binding site geometrically, it must also match the type of interaction belonging to the part of the binding site. For example, if region 2 on the molecule is hydrophobic, then it must be again matched by the hydrophobic region in the binding site. Any other interaction will have a positive free-energy change, which means repulsion. Thus a "mismatch" may actually provide a negative contribution to the binding and *enhance* the overall recognition.

Another factor affecting the selectivity is the entropy contribution to binding. We see from equation (A-13) that the increase in entropy favors binding. This is the basis of probably the most important bond in biological systems, the hydrophobic bond. Figure 1-2 depicts the situation as it exists in a vacuum, obviously not a "typical biological situation." The biological sensory systems work in an aqueous environment. This means that water must be included in the thermodynamical considerations (Figure 1-2b). Water molecules interact with both S and X because its hydrogen bonds, and ion–dipole and dipole–dipole interactions have enough energy to cause such an association. This phenomenon is, of course, called *hydration*. When S and X are separated (Figure 1-2b) there are (arbitrarily) two free (random) molecules of water present in the system. For a unique description of the system we must specify their coordinates. If the origin of the coordinate system is located in the binding site, then the position of the remaining particles is described uniquely by $3n = 9$ coordinates. Those are their degrees of freedom. On the other hand, when complex SX is formed (Figure 1-2c) thirteen more water molecules will have been liberated from the cleft, thus increasing the number of particles to 15 and the number of degrees of freedom to 45 (the numbers correspond, of course, to Figure 1-3 but are otherwise arbitrary). This leads to an increase in the entropy (lowering the free energy) of the whole system. Thus the binding free energy is driven by the entropy (of water), which is the basis of the *hydrophobic bond*. Its equivalent enthalpy value/interaction is included in Table 1-1 for comparison.

For the purpose of this discussion it is most important to realize that the hydrophobic bond will contribute to the free energy of interaction *only if the two molecules fit geometrically and eliminate some hydration water from the binding cleft and its vicinity*. It is this condition which accounts for the enormous selectivity found in immunochemical reactions, biological receptor binding, enzyme/substrate recognition, etc. It is also combined with enthalpy-driven binding, which can again act only at a relatively short range (cf. Table 1-1).

Equilibrium constants of some immunochemical reactions have a stronger dependence on temperature than others, implying that the relative contribution to free-energy change from, e.g., the hydrogen bond (enthalpy) and the hydrophobic bond varies.[3] The important corollary for the design of so-called biosensors is that it would be difficult to employ a water-based biological selective system (e.g., most enzymes or antibodies) for nonaqueous applications such as gas sensing.

The discussion of selectivity has so far been limited to thermodynamics. Selectivity can also arise as a result of *kinetics* (see Appendix B). One example of this type is the extraction of ions which have different mobilities inside the ion-selective membrane layer; the membrane interacts preferentially with the faster ion. It can be also seen in the Michaelis–Menten mechanism of enzymatic reactions in which the shape-recognition (thermodynamic) selectivity is determined by the formation of the enzyme–substrate complex ES (cf. Section 1.2.2).

1.2.1. Evaluation of Selectivity

Many different species which are present in the sample contribute to the output signal of a sensor. High selectivity means that the contribution from the *primary species* X dominates and that the contribution from the *interfering* species I_i is minimal. This statement can be expressed in the form of a general equation relating the sensor output S_{out} to the composition of the sample:

$$S_{out} = \Re \left(a_x^n, \sum_i K_{i,x} a_i^m \right) \tag{1-22}$$

where $\Re$ is the general response function defined in equation (1-9) for a single species (x) interaction. The activities of the primary species a_x and of the interfering species a_i are raised to powers n and m, respectively. The constant $K_{i,x}$ is called the *selectivity coefficient*. It is pertinent to note here that the origin of the selectivity coefficients and of the activity exponents lies in the *mechanism* of the interaction between all the species and the sensor. Their study provides the most valuable insight into the details of the operation of any sensor.

It follows from equation (1-22) that an *ideal sensor* would respond *only* to the primary species, and therefore that all selectivity coefficients should be zero for all interfering species. Clearly, this is not the case in reality and one of the most important functions in evaluating the performance of a chemical sensor is the quantification of its selectivity. For one of the most advanced chemical sensors, ion-selective electrodes, it has been customary to study and evaluate the potentiometric selectivity coefficients $K_{i,x}$ by using the specific form of equation (1-22), the so-called Nikolskij–Eisenman equation,

$$E_{out} = RT/z_x F \left[\ln \left(a_x + \sum_i K_{x,i} a_i^{z_x/z_i} \right) \right] \tag{1-23}$$

In this case the activity of the primary ion is designated as a_i, while z_x and z_i are the charges of the primary and interfering ions, respectively.

There are two operating regions of the sensor as defined by the relative value of the two terms in the logarithmic argument in equation (1-23), illustrated in Figure 1-4. If the primary ion activity is much greater than the second term, equation (1-23) reduces to the normal form of the Nernst equation (cf. Section 4.1.1) and the sensor responds only to the change in activity of the primary ion with the slope of the response equal to $RT/z_j F$,

$$E_{out}(A) = RT/z_x F \ln a_x \tag{1-24}$$

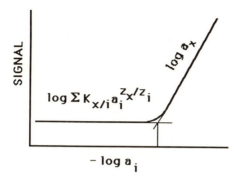

Figure 1-4. Evaluation of the selectivity coefficient K_{ij} of a logarithmic sensor.

On the other hand, if the activity of the primary ion is small compared to the second term, equation (1-23) becomes

$$E_{out}(B) = RT/F \ln \left(\sum_i K_{x,i} a_i^{1/z_i} \right) \qquad (1\text{-}25)$$

If we prepare a solution which contains only the primary and one interfering ion, in principle we can evaluate the selectivity coefficient for that particular ion by solving equations (1-24) and (1-25):

$$K_{x,i} = \frac{a_x^{1/z_x}}{a_i^{1/z_i}} \qquad (1\text{-}26)$$

The selectivity coefficient here indicates that the sensor prefers the primary ion relative to the interfering ion. The smaller its value, the higher the preference for the primary ion and the more selective the sensor.

The shape of the curve shown in Figure 1-4 is typical of sensors whose response function $\Re$ is logarithmic. This approach can be applied to other sensors with a different relationship linking the output to the activity of the species in the sample. Thus, e.g., for an amperometric or some optical sensors, a linear response function $\Re$ would be more appropriate,

$$S_{out} = s_x(a_x + S_i K_{x,i} a_j) \qquad (1\text{-}27)$$

In this case the graphic representation of the response is shown in Figure 1-5. From the above argument about the relative contribution of the two terms on the right-hand side of equation (1-27), we can see that the selectivity coefficient for linear sensors is simply the ratio of the intercept

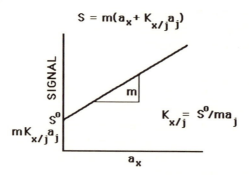

Figure 1-5. Evaluation of the selectivity coefficient of a linear sensor.

of the response curve S^0_{out} to the activity of the interferants, normalized with respect to the slope of the primary species s_j,

$$K_{x,i} = S^0_{out}/s_i s_x a_i \qquad (1-28)$$

This method obviously fails for sensors whose response function does not have a linear dynamic range, such as potentiometric enzyme sensors. In that case the only alternative in reporting the selectivity of the sensor seems to be to present the entire calibration curve for each set of experimental conditions under which the response has been obtained.

It is common practice to evaluate the response of a sensor only for increasing values of the activity (concentration) of the primary species. This is mainly due to the fact that it is more convenient to prepare continuously the broad range of the test concentrations by adding increasing amounts of a concentrated (or pure) primary species to the background sample than *vice versa*. This is particularly true if the sensor responds in the range of concentrations where the background solvent or gas is in a large molar excess. In that case, in order to decrease the concentration of the primary species by an order of magnitude in a volume of sample V_0 it is necessary to add $9V_0$ of the background solution or gas. This practice has some dangers, particularly if the sensor does not behave reversibly. An excellent way to overcome this problem is to test the sensor in a flowing stream of the background sample and to inject the primary compound. Depending on the volume of the injected sample the concentration change can be described as a step or as an impulse, the difference being in the width of the plateau. In that case the sensor response can be characterized on both the ascending and descending edges of the impulse. For liquid samples the flow injection analysis (FIA) offers the best characterized test conditions (cf. Section 1.3).

An absolutely selective sensor really does not exist; there is always some interference present. Another approach by which true information can be obtained from less than ideal sensors is to use them in an array and evaluate their response statistically.[4]

1.2.2. Enzymatic Layers

Enzyme-containing layers have been used as a means of achieving selectivity in all types of chemical sensors. The specific differences which arise from the various operating principles of different sensors will be discussed separately. Here we focus only on enzymatic reactions.

Enzymes are a special kind of catalyst, proteins of molecular weight 6–400 kdaltons found in living matter. They have two remarkable properties: (1) they are extremely selective to a given substrate and (2) they are extraordinarily effective in increasing the rates of reactions. They therefore combine the recognition and amplification steps. A general enzymatically catalyzed reaction can be described by the Michaelis–Menten mechanism in which E is the enzyme, S the substrate, and P the product:

$$S + E \underset{k_1}{\overset{k_1}{\rightleftharpoons}} ES \overset{k_2}{\Longrightarrow} P + E \qquad (1\text{-}29)$$

The reaction velocity v can be expressed as the rate of increase in the concentration of the product,

$$v = dC_P/dt = k_2 C_{ES} \qquad (1\text{-}30)$$

For a high value of the substrate concentration, the reaction velocity attains its maximum. Under those conditions all the available enzyme E_T is bound in the complex with the substrate. Hence

$$v_{max} = k_2 C_{ET} \qquad (1\text{-}31)$$

This means that the maximum velocity is proportional to the concentration of the enzyme. The enzyme is present in this reaction either free or complexed with the substrate,

$$C_{ET} = C_E + C_{ES} \qquad (1\text{-}32)$$

At steady state the concentration of the enzyme complex is constant,

$$dC_{ES}/dt = k_1 C_S C_E - (k_{-1} + k_2) C_{ES} = 0 \qquad (1\text{-}33)$$

In this case the Michaelis–Menten constant K_m is defined by

$$K_m = (k_{-1} + k_2)/k_1 = C_S C_E/C_{ES} \qquad (1\text{-}34)$$

Substitution for quantity C_E in equation (1-34) from equation (1-32) yields

$$K_m = C_S(C_{ET} - C_{ES})/C_{ES} \qquad (1\text{-}35)$$

which, when combined with equations (1-30) and (1-31), gives

$$K_m = C_S(v_{max} - v)/v \qquad (1\text{-}36)$$

Rearrangement leads to the well-known Michaelis–Menten equation

$$v = v_{max} C_S/(C_S + K_m) \qquad (1\text{-}37)$$

It can be shown that K_m equals the concentration of the substrate at which the reaction velocity is one-half its maximum. It is a characteristic constant for a given enzyme.

The activation energy is decreased (i.e., the rate of reaction is increased) by *catalysis*. The extraordinary specificity of enzymatic catalysis is accounted for by the entropy term in equation (B-7). It is achieved physically by the enzyme (a protein) having a stereospecific *binding site* in which the two reactants can be brought together at the protein surface in a precise and favorable orientation for the overall reaction to take place.

Like any other proteins, enzymes are subject to acid–base equilibria which affect their catalytic properties, i.e., their values of K_m. Each enzyme has its own characteristic pH dependence $\Re(pH)$. Thus the general Michaelis–Menten equation which takes into account this pH dependence of K_m is

$$v = v_{max} C_S/\Re(pH)(C_S + K_m) \qquad (1\text{-}38)$$

Besides hydrogen ions, other species also affect the enzymatic catalytic activity. This phenomenon is called *inhibition* and may be *specific, nonspecific, reversible,* or *irreversible*.

Enzymatic reactions combine substrate specificity with high amplification factor. From that viewpoint they are ideal selective layers for chemical sensors. However, they are not specifically part of the information acquisition–processing scheme in Nature. Their exclusive role is to lower, in a highly selective manner, the activation energy barrier of certain reactions, thus acting as a regulator.

A schematic diagram of an enzymatically coupled chemical sensor is shown in Figure 1-6. The geometry of the probe is chosen such that it corresponds to a semi-infinite diffusion. Other, for instance radial, geometries have been considered. The sensing element can be a heat probe, an electrochemical sensor, or an optical sensor. Mass sensors are generally unsuitable, because they do not perform well in a condensed medium (i.e., water) and partly because enzymatic reactions are *catalytic* reactions in which the mass difference between the total reactants and the total products is not expected to be significant. The basic operating principle is simple: an enzyme (a catalyst) is immobilized inside a layer into which the substrate(s) diffuses. Hence it reacts according to the Michaelis–Menten mechanism and the product(s) diffuses out of the layer into the sample solution. Any other species which participate in the reaction must also diffuse in and out of the layer. Owing to the combined mass transport and chemical reaction this problem is often referred to as the *diffusion-reaction* mechanism. It is quite common in electrochemical reactions where the electroactive species, which undergo chemical transformation at the electrode, participate in some coupled chemical reactions. This case is described mathematically by a set of higher-order partial differential equations, which are usually solved numerically. Enzymes operate only in an aqueous environment, so the immobilization matrices are gels, specifically hydrogels.

The general diffusion-reaction equation for species i in one direction (x) is

$$\partial C_i/\partial t = +D_i(\partial^2 C_i/\partial x^2) \pm R(C_i) \tag{1-39}$$

When the pH-dependent Michaelis–Menten equation (1-38) is substituted for the reaction term $R(C_i)$, we obtain for substrate S the equation

$$\frac{\partial C_S}{\partial t} = D_S \frac{\partial^2 C_S}{\partial x^2} - \frac{V_m C_S}{\Re(\text{pH})(C_S + K_m)} \tag{1-40}$$

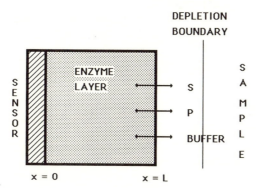

Figure 1-6. Schematic diagram of an enzymatic sensor.

It is convenient to normalize the variables in equation (1-40) as follows:

$$t = t L^2/D, \qquad C_S = C_S K_S, \qquad x = x L \tag{1-41}$$

where t, C_S, and x are dimensionless variables, and ϕ is the Thiele modulus given by

$$\phi = L[V_m/K_m D_S \Re(\text{pH})]^{1/2} \tag{1-42}$$

Equation (1-40) then assumes the form

$$\frac{\partial C_S}{\partial t} = D_S \frac{\partial^2 C_S}{\partial x^2} - \frac{\phi^2 C_S}{1 + C_S} \tag{1-43}$$

For $\phi > 10$ the mechanism is diffusion controlled. Conversely, for a small value of the Thiele modulus the process is reaction controlled, which means that the diffusional fluxes of all species participating in the reaction governed by equation (1-43) are greater than the reaction term. This transformation is carried out for all variables. The actual solution of this problem depends on the initial and boundary conditions. These, in turn, depend on the approximations chosen for the model.

Let us now review briefly the approximations that have been made by various workers for the enzymatically coupled sensors and rank them in some approximate order:

1. Linear diffusion gradient inside the enzyme layer.
2. No partitioning of reactants and products between the gel and the sample.
3. No depletion layer at the gel/sample boundary.
4. No effect of mobile buffer capacity.
5. No effect of fixed buffer capacity.
6. No pH dependence of K_m.
7. No Donnan potential at the gel/sample boundary.
8. Rates of protonation reactions are fast.

Approximation (1) is bad: the concentration profiles are not linear. Partitioning of species between the gel and the sample (approximation 2) is also related to the existence of the Donnan potential (7), but it is a problem even for electrically neutral species (such as oxygen). If the solution is stirred, the effect of the depletion layer at the gel/membrane interface is negligible (3). However, it could be a problem in stationary solutions. Approximations (4) through (6) would be the most serious for enzymatic sensors in which the output is related to the change of pH, because then the buffer capacity would have to be low. However, for sensors which use some other reactants/products *but hydrogen ion*, a large excess of buffer would make the effects due to these assumptions negligible. Finally, the rates of (almost) all protonation reactions (8) are extremely fast and can be safely assumed to be instantaneous on the time scale of all the other processes.

With these assumptions in mind we now complete the outline of the solution of the diffusion-reaction problem as it applies to the most difficult case, the pH-based enzymatic sensors. We assume only that there is no depletion layer at the gel/solution boundary (3), no fixed buffer capacity (5), and that the protonation reactions are very fast (8). The objective of this exercise is to determine the optimum thickness of the gel layer, which is critically important for all zero-flux-boundary sensors, as follows from equation (1-42).

As a rule, the hydrogen ion is involved not only in the pH dependency of the reaction term (the Thiele modulus), but also as the actively participating species involved in the acid–base equilibrium of the substrates, reaction intermediates, and products. Furthermore, enzymatic reactions are always carried out in the presence of a mobile buffer. By mobile we mean a weak acid or a weak base, which can move in and out of the reaction layer, as opposed to an immobile buffer represented by the gel (and the protein) itself. Consequently, we must include the normalized diffusion-reaction equation for hydrogen ion and for the buffer:

$$\frac{(\partial C_{H+})_{tot}}{\partial t} = D_{H+} \frac{\partial^2 C_{H+}}{\partial x^2} - \frac{\phi^2 C_S}{(1+C_S)} + D_{HA} \frac{\partial^2 C_{HA}}{\partial x^2} \qquad (1\text{-}44)$$

where $(C_{H+})_{tot}/\partial t$ is the change in the total (bound and unbound) protons within the enzyme layer, while the third term is the flux of the buffer acid. For simplicity we consider here a simple monoprotic buffer. These two equations have to be coupled with the buffer dissociation equilibrium,

$$K_a = C_{H+} C_{A-}/C_{HA} \qquad (1\text{-}45)$$

Next we define the boundary and initial conditions. For "zero-flux" sensors there is no transport of any of the participating species across the sensor–enzyme layer boundary. Such a condition would apply to, e.g., optical, thermal, or potentiometric enzyme sensors. In that case the first space derivatives of all variables at point x are zero:

$$\{C_S(0, t)\}'_x = \{C_{H+}(0, t)\}'_x = \{C_{HA}(0, t)\}'_x = 0 \qquad (1\text{-}46)$$

On the other hand, amperometric sensors would fall into the category of "nonzero-flux sensors" by this definition and the flux of at least one of the species (product or substrate) would be given by the current through the electrode.

The lack of depletion layer at the gel–solution boundary $(x = L)$ is guaranteed by stirring the sample. Under those conditions the concentration of all species at that boundary are equal to the bulk values:

$$C_S(L, t) = C_{S,\text{bulk}}, \qquad C_{HA}(L, t) = C_{HA,\text{bulk}}$$

$$C_H(L, t) = C_{H,\text{bulk}}, \qquad C_A(L, t) = C_{A,\text{bulk}} \qquad (1\text{-}47)$$

The initial conditions are

$$C_{HA}(x, 0) = C_{HA,\text{bulk}}, \qquad C_A(x, 0) = C_{A,\text{bulk}}, \qquad C_H(x, 0) = C_{H,\text{bulk}}$$

$$C_S(x, 0) = 0 \qquad \text{for} \quad x < L \qquad (1\text{-}48)$$

which means that all species except the substrate are initially present inside the enzyme layer.

This is a system of stiff, second-order partial differential equations which can be solved numerically to yield both transient and steady-state concentration profiles within the layer. Because the concentration profile changes most rapidly near the $x = L$ boundary, an ordinary, finite-difference method does not yield a stable solution and is not applicable. Instead, it is necessary to transform the distance variable x into a dummy variable y using the relationship

$$y = L[1 - \exp(-ax) + x \exp(-a)] \qquad (1\text{-}49)$$

This transformation allows for equal distribution in y-space while concentrating the lines close to the $x = L$ boundary. Parameter a sets the spacing of the lines.[5] The above technique is called MOL1D (Method Of Lines in 1 Dimension) and is suitable for solving parabolic and hyperbolic initial boundary-value problems in one dimension. It employs a finite-differences method to approximate the space derivatives and solves the resulting system of ordinary differential equations by using either the Adams–Bashford–Moulton or Gear method for stiff equations.[5]

The actual solution for both *transient* and *steady-state* response of any zero-flux boundary sensor can be obtained by solving equations (1-43) through (1-49) for the appropriate boundary and initial conditions. Comparison of the experimental calibration curves and of the time response curves with the calculated ones provides verification of the proposed model. It is hence possible to determine the optimum thickness of the enzyme layer. The Thiele modulus is the controlling parameter in the diffusion-reaction equation, so it is evident from equation (1-42) that the optimum thickness L will depend on the other constants and functions included in the Thiele modulus. For this reason the optimum thickness will vary from one kinetic scheme to another.

An additional important observation is related to the *detection limit*, *dynamic range*, and *sensitivity*. For the expected values of the diffusion coefficient (in the gel) of approximately 10^{-6} cm^2 s^{-1} and substrate molecular weights about 300, the detection limit is approximately 10^{-4} M. This is due to the fact that the product of the enzymatic reaction is being removed from the membrane by diffusion at approximately the same rate as it is being supplied. The dynamic range of the sensor depends on the values of quantities K_m and V_m (which depend on the enzyme loading). Generally speaking, higher loading should extend the dynamic range at the top concentration values. It is often stated incorrectly that "the enzyme sensor has close to *theoretical* dependence" or a "nernstian response," which means that a one-decade change in the *bulk* concentration of the substrate is expected to yield a one-decade change in the *surface* ($x = 0$) concentration. In the case of potentiometric enzyme sensors it would yield a slope of approximately 60 mV per decade. It is not intuitively obvious but clearly evident from comparison of the experimental and calculated response curves that there is no general theoretical slope, each enzymatic sensor having its own value depending on the mechanism and on the conditions under which it operates. We must remember that a decade/decade slope would occur only if a *constant fraction* of the product were to reach the $x = 0$ interface. The upper limit of the dynamic range depends on the value of the ratio $V_m/K_m D_S \Re(\text{pH})$ in the Thiele modulus. It can be increased by enzyme loading but, obviously, only up to a point. Normally the dynamic range is approximately between 10^{-4} and 10^{-1} M.

1.2.3. Immunochemical Selectivity

If one were to define an ideal reagent for the construction of a selective layer, antibodies would have to be considered very seriously. Their selectivity is based on stereospecificity of the binding site for the antigenic determinand (antigen, hapten, epitope). Their production is relatively inexpensive and universal, which means that an antibody for any antigen, regardless of its shape or chemical nature, can be produced by the same general procedure. In this respect the only limitation seems to be that of size: antigens of molecular weight less than 2000 daltons normally do not induce an immunochemical response in B lymphocytes which produce them. In order to obtain antibodies for low-molecular-weight antigens (haptents), it is necessary to link the latter to a high-molecular-weight polymeric carrier (such as bovine serum albumin or polyethylene glycol).

Antibodies belong to the group of serum proteins called immuno-globulins.[6] Their molecular weight ranges from 140 kD to 970 kD. The number of antigens that can be bound to one antibody determines their *valency*, which is typically 2 but can be as high as 10 for immunoglobulin M. Their primary function is to disable foreign (high molecular weight) immunogens, be they proteins, nucleic acids, viruses, etc., which may invade the organism. In that respect they can be regarded as highly specific complexing agents, which form one of the key factors in the defense mechanism. The most common antibody is immunoglobulin G (IgG), which has molecular weight 146 kD and valency 2 (Figure 1-7). The diameter of the molecule is estimated to be about 70 Å and the area of the binding site varies between 250 and 400 Å^2. Traditionally, it has been produced by multiple sensitization of an animal (such as a rabbit, goat, or dog) to a suitable antigen. When the natural immune response triggers the production of IgG targeted against the immunogen, IgG can be isolated from the animal antiserum and purified to the desired level. Antibody produced by this procedure is called *polyclonal*. A more recent procedure involves fusion of the sensitized B lymphocytes with a myelloma cell (malignant cancer cell) preparation, which is then implanted in the animal.

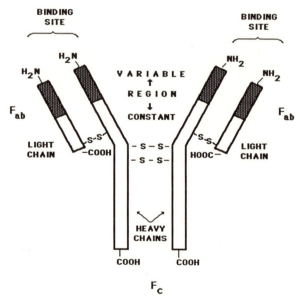

Figure 1-7. Double-valent antibody molecule. IgG can be cleaved in the ($-S-S-$) hinge region yielding one F_c and two F_{ab} fragments which retain their binding properties.

It produces a large amount of specific *monoclonal* antibodies, which are harvested and purified.

Although the active site on the antibody is fundamentally highly specific to the given antigen, any preparation of antibodies, either poly- or monoclonal, is heterogeneous. This heterogeneity is far greater in polyclonal antibodies than in their monoclonal counterparts. The *average* ability to complex antigen is called the *avidity* of the preparation, while the binding equilibrium between an antibody (Ab) and an antigen (Ag) is referred to as the *affinity*:

$$\text{Ab} + \text{Ag} \xrightleftharpoons[k_r]{k_f} [\text{Ab.Ag}] \qquad (1\text{-}50)$$

The binding constant K is defined as the ratio of the forward and reverse rate constants [see equation (1-50)]. It ranges from $K < 10^4$ liter mol^{-1} for weak binding of antigens of molecular weight less than 2500 D, to $K > 10^9$ liter mol^{-1} for antigens of molecular weight greater than 6–8×10^6 D which are very strongly bound.[2]

Under physiological conditions the equilibrium constant values in the range $K = 10^5$ to 10^9 liter mol^{-1} correspond to -6 to -11 kcal mol^{-1} of the standard free energy. The forward rate of the immunochemical reaction is invariably very high (diffusion limited). This is consistent with the strategy of the biological defense mechanism: "shoot first, ask questions later," where the inactivation of a "potentially harmful" antigen must be conducted with maximum speed but the recognition of a truly harmful (or innocuous) constituent can be undertaken much more slowly. This means that the dissociation (k_r) rate constants vary over 7 decades from 10^{-4} s^{-1} to 10^3 s and determine the overall affinity of the hapten or antigen to the antibody.

The nature of the Ab−Ag bond is of critical importance for chemical sensing.[2,7] The most prevalent bonds are considered to be coulombic and van der Waals interactions. The role of water in the overall binding is critically important. First of all it is the prerequisite in the formation of the hydrophobic bond. However, expulsion of water from the binding site, which takes place during binding, decreases the local dielectric constant and increases the strength of the coulombic and van der Waals bonds (Table 1-1) in that region. Thus the binding is cooperative. The close stereospecific fit is, of course, necessary. There is no covalent bonding involved in any immunochemical reactions.

The binding equilibrium expressed above in equation (1-50) is a gross oversimplification of the situation. The heterogeneity of the binding sites and the multiple valency of individual antibodies lead to the formation of secondary bonds,[2] which contribute to *hysteresis* or "ripening" of the complex. Its ultimate form is the polymerization of primary complex which happens when the antigen is also polyvalent. Formation of the polymer (precipitin reaction) renders the immunochemical reaction virtually irreversible.

The secondary bonds, which may be formed much more slowly than the primary bonds, actually contribute more to the overall affinity. For example, the primary (coulombic) bond between bovine serum albumin (BSA) and anti-BSA IgG is $3.3 \, \text{kcal mol}^{-1}$ while the secondary bond (van der Waals) is $6.8 \, \text{kcal mol}^{-1}$ for the total $K = 10.1 \, \text{kcal mol}^{-1}$. The formation of the secondary bond is much slower, so it is easier to *prevent* formation of the strong complex rather than to try to *dissociate* it. This is one reason why the *competitive* immunoassays yield results that correlate with the equilibrium binding constants, but any *direct* binding assays must rely on measurement of the initial rate of binding.

In order to assess the utility of the immunochemical reaction for chemical sensing, we need to examine the effects of the experimental conditions on the primary association reaction. The effect of temperature is not particularly distinct for most reactions and cannot be generalized. This is due to the fact that the relative contribution of the hydrophobic (entropic) bond and other (enthapic) bonds is different. The equilibrium is largely insensitive to the effects of pH (6.5–8.5) and ionic strength. The presence of organic solvents begins to play a role only when the hydrophobic bonds become affected. Obviously, the presence of water in the binding process is mandatory.

The dissociation of the "aged" immunochemical complex can be achieved, but sometimes under denaturing conditions. Techniques which have been used mainly for preparative purposes include the use of chaotropic salts (such as KCNS, tetraethylammonium chloride, and guanidine hydrochloride). These salts compete for "available" water. Ionic (sodium dodecyl sulfate) and nonionic (polyethyleneglycol) detergents as well as solvents (ethanol, dimethyl sulfoxide) play a similar role. In general, these agents tend to disrupt the hydrophobic bond and must be applied in relatively high ($\sim$ M) concentrations. Disruption of hydrogen bonds, leading to denaturation in most cases, is done by addition of 6–8 M urea and/or by lowering the pH to approximately 2. It has been possible to dissociate charged antigen by placing the complex in a high electric field. This is called *electrophoretic dissociation*, which works only in solutions of very low ionic strength. Finally, it is sometimes possible to use low-molecular-weight haptens, particularly in the early stages of complex formation. In that sense, dissociation with the help of hapten can be regarded as competitive binding.

Are antigens really ideal binding sites for sensor applications? The answer to this question is unclear at present. First, the high (and time-variable) value of the affinity binding constant of (particularly) polyclonal antbodies makes the interaction virtually irreversible. We have to combine this with the fact that the IgG molecule is large and that the area occupied by one active site is also very large. This means that the packing density is very low and the dynamic range [see equation (1-8)] is narrow. This situation can be somewhat improved by isolating the binding sites (i.e., the F_{ab} fragments) from the rest of the molecule. Furthermore, we must eliminate the multivalency in order to prevent polymerization. Again, this can be achieved at the level of a single F_{ab} fragment. The cost of preparing single F_{ab} fragments is, however, much higher. A good strategy for improving the dynamic range would be to increase the number of binding sites by immobilizing them in a layer, rather than on the surface, and to incorporate sites with very different values of the binding constant. The formation of the secondary bonds should be prevented. There is no simple recipe for doing this, except a mode of operation resembling competitive binding. The presence of water is mandatory for the formation of the complex.

The above factors clearly place substantial constraints on the types of chemical sensor which could utilize immunochemical selectivity. We can exclude thermal sensors, because the absolute amount of heat evolved during a single binding step would be too small to measure. Assuming that the sensor would be operated continuously in water, mass sensors would be possible but only if the effects of increased mass due to the immunochemical reaction were not obliterated by other nonspecific effects at the sensor–water interface (cf. Section 3.2). However, the number of available binding sites may be a limiting factor. There is no presently known mechanism by which the immunochemical reaction could be utilized electrochemically for direct sensing. The best possibility seems to be the optical mode of sensing, where some very promising preliminary explorations have already been made.

In contrast, sensor systems seem to be ideal for the examination of immunochemical selectivity. Incorporation of manipulative step(s) opens the door for regeneration of the antibody, or even for operation under equilibrium conditions. Thermal, electrochemical, optical, and mass immunochemical sensor systems have been demonstrated.

1.2.4. Protein Immobilization

The choice of the type of immobilization procedure is largely dictated by the intended application. In some cases we may wish to have a thick layer of protein, in others it is preferable to have only surface immobilized sites. The first condition of immobilization is that it must be gentle, because the protein activity must be preserved. A physical entrapment using a dialysis membrane, or gel entrapment, covalent attachment to gel matrix, or direct covalent bonding to the surface of the sensor have been used. A comprehensive review of various immobilization procedures has been compiled.[8] Another approach is to use whole bacterial cultures, whole tissues, or even organs. In that case the advantage is that the enzyme is in its natural, protected environment and lasts longer.[9] The problem is that the response is usually very long, minutes to hours.

The simplest method of enzyme immobilization, and the one used most often, is co-crosslinking of the enzyme with some inert protein, such as collagen, using glutaraldehyde:

$$collagen - NH_2 + COH - (CH_2)_3 - COH + H_2N - enzyme$$

$$\Rightarrow collagen - N = CH - (CH_2)_3 - CH = N - enzyme$$

This step is sometimes followed by reduction of the amide bonds to yield more stable amino-linkages:

$$collagen - N = CH - (CH_2)_3 - CH = N - enzyme$$

$$\xrightarrow{\text{LiBH}_4} collagen - NII - CH_2 - (CH_2)_3 - CH_2 - NH \quad enzyme$$

Other matrix proteins, or polymers such as albumin or poly(vinyl amine), can be also used. Preferably, the final gel must be uncharged at the normal operating range of pH (4–9) in order to avoid the Donnan potential (ion-exchange potential) at the gel–sample interface, to minimize osmotic swelling, and to minimize buffering by the gel. For these reasons uncharged gels are preferable. A simple procedure for preparing such an uncharged, enzyme-containing gel uses polyacrylamide, which does not contain charged groups in the normal pH operating range of most enzymes.[10] In this case the enzyme is derivatized using N-succimidyl methacrylate (NSM):

The activity of the derivatized enzyme (penicillinase, peroxidase, and glucose oxidase) is unaffected. It is then co-polymerized with acrylamide, bis-acrylamide, and N,N,N',N'-tetramethylene diamine using ammonium persulphate as initiator:

$$CH_2=\overset{\overset{\displaystyle CH_3}{|}}{C}-\underset{\underset{\displaystyle O}{\|}}{C}-NH-\boxed{P}\ +CH_2=CH-\underset{\underset{\displaystyle O}{\|}}{C}-NH_2$$

$$\longrightarrow [(CH_2-CHCONH_2)_n-\overset{\overset{\displaystyle CH_3}{|}}{C}HC-\underset{\underset{\displaystyle \boxed{P}}{|}}{C}ONH]_m$$

The result is the enzyme covalently linked to an unchanged gel matrix. For fabrication of integrated circuit sensors, it is highly desirable to be able to deposit and pattern the enzyme layer using photolithography. Enzyme-containing layers have been prepared using this procedure,[11] but so far only in limited thicknesses ($< 10\ \mu m$).

The enzyme-containing layer (gel) represents a relatively concentrated medium. In fact it can be regarded as a separate phase. Both charged and uncharged solutes partition between the sample solution and the enzyme-containing gel.[10] Typical values of the partitioning coefficient vary between 0.4 and 1; it depends both on the charge and polarity of the permeating species. This fact, which is very important for the prediction and modeling of the response of enzyme sensors of any kind, has so far been mostly ignored.

The usable lifetime of the enzyme layers depends on the ruggedness of the enzyme and on the matrix in which it is immobilized. For oxidases, the continuous-use lifetime in the polyacrylamide matrix is approximately two weeks, while in the polyvinylalcohol matrix it is several months.[12] The limiting factors are usually the irreversible inhibitors, such as hydrogen peroxide or heavy metal ions.

The choice of method for the immobilization of proteins at surfaces is dictated above all by the nature of the surface itself, namely, by the availability of the reactive groups at that surface. Most common is the hydroxyl group which is found on oxides, carbohydrates, and most glasses. The most general reaction used for immobilization via this group is the silanization

$$\begin{bmatrix} -OH & Cl \\ -OH \\ -OH & Cl \end{bmatrix} + Cl-Si-(CH_2)_n-CH_2-X \Rightarrow \begin{bmatrix} -O \\ -O-Si-(CH_2)_n-CH_2-X \\ -O \end{bmatrix}$$

Table 1-2. Typical Immobilization Reactions[a]

Reactive group I (e.g., on surface)	Intermediate	Reactive group II (e.g., on terminal reagent)	Coupling linkage "type"
$-R-NH_2$	$R'-N=C=N-R''$ (Carbodiimide)	$HO-\overset{\overset{O}{\|}}{C}-$	$-NH-\overset{\overset{O}{\|}}{C}-$ "Amide"
⟨○⟩$-NH_2$	$-$⟨○⟩$-\overset{\oplus}{N_2}\overset{\ominus}{Cl}$ "Diazonium"	$HO-$⟨○⟩$-$	$-$⟨○⟩$-N=N-$⟨○⟩ "Diazo" HO
⟨○⟩$-NH-NH_2$		$-\overset{\overset{O}{\|}}{C}-$	⟨○⟩$-NH-N=C-$ "Hydrazide"
$-SH$		$HS-$	$-S-S-$ "Disulfide"
"Maleic anhyride"		H_2N-	⟨COOH / CONH-⟩ "Amide"
$-COOH$	$-\overset{\overset{O}{\|}}{C}-CN_3$	H_2N-	$-CONH-$ "Amide"
$-CHO$	"Schiff base"	H_2N-	$-C=N-$
$-SH$	$R'-N=C=N-R''$	$HO-\overset{\overset{O}{\|}}{C}-$	$-S-CO-$ "Thioamide"
⟨OH / OH⟩	CNBr	H_2N-	⟨OCO-NH- / OH⟩
cyanuric Cl	"Cyanuric chloride"	H_2N-	cyanuric NH-

This reaction introduces the reactive group $X_|$ (e.g., NH_2, SH, epoxy), which is suitably "spaced" from the surface by the n CH_2 groups. The trichlorosilane is called a *coupling agent*. There are many different coupling agents from which to choose (Table 1-2). The mechanical and chemical stability is usually taken for granted but, in fact, it can be the most significant factor in the gradual loss of the immobilized moiety. For hydrophobic surfaces used in aqueous environment one can resort to a generally applicable immobilization procedure, which relies on the entrapment of the hydrophobic chain of the coupling agent:

$$\Big|+ \text{--------}X \xrightarrow{\text{swelling}} \text{----}\Big|\text{--} X \xrightarrow{\text{de-swelling}} \text{----}\Big|\text{--} X$$

It should be noted that the straight lines representing the surfaces in the above diagrams do not imply that the surface is "smooth on the angstrom scale."

1.3. Systems

In discussing the types and operating principles of various sensors we must remember that they are only part of the total *information acquisition* process. In some cases it is necessary to perform some additional operations, such as sampling, sample manipulation, additions of various reagents, incubation, etc. Those steps may or may not involve the sensor itself, yet they are done in order to gain some specific advantage. The data are then obtained in *discrete steps* rather than *continuously*. We shall call this a *systems* approach and in this section we show that there are advantages and disadvantages in using the sensor in this way. The choice between the discrete or continuous mode must be governed by considerations which depend on the requirements of the measuring situation. The line between the system and the sensor itself is often blurred.

Figure 1-8 shows a record of a signal from an arbitrary sensor on an

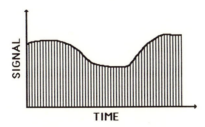

Figure 1-8. Continuous vs. discrete information acquisition.

arbitrary time scale. Without any more information we cannot tell if it was obtained in a discrete or in a continuous fashion, but this uncertainty does not diminish its information content. It could have been obtained as a continuous recording of the sensor output with the ordinate lines drawn later; alternatively, each ordinate could represent the experimental point obtained as the result of a discrete operation with the points subsequently connected as an envelope. The important thing to realize is that, in such a situation, an injection *of known concentration* could have been introduced repeatedly anywhere in this sequence and would *calibrate* the system intermittently. In order to ensure that no information was lost it is necessary to know that the *information acquisition frequency* was greater than the frequency of the fastest detectable event.

Obviously, it would be ideal to record the state of the variable of interest continuously, because then we *know* that we have obtained all the information, but that would require an *ideal sensor* which would possess long-term calibration stability, stable sensitivity, selectivity, and dynamic range.

By implementing the discrete mode of measurement we obtain an additional degree of freedom to arrange the experiment such that the weak points of a sensor are de-emphasized and its strong points enhanced. For this reason almost every type of chemical sensor has been used in a system configuration.

The hardware necessary for those operations includes solenoid valves, pumps, fluidic switches, injection valves, and other moving parts, usually all controlled by the same microprocessor. There are also passive elements, such as connecting and delay lines, mixing, dialysis and permeation chambers, stream splitters, and phase separators. All these elements are arranged so that certain repetitive operations can be performed precisely and in a correct sequence. Such systems have been used for many decades and are known as automatic analyzers. A discussion of their principles and operational characteristics is clearly beyond the scope of this book. Yet, they are so intimately coupled with chemical sensors that it is often impossible to draw a clear line between a "pure" sensor and a "system containing a sensor." As the systems become smaller and smaller, this line becomes more difficult to define owing to modern microfabrication techniques.

1.3.1. Flow Injection Analysis

One of the most elegant and efficient ways of manipulating analytical amounts of liquid samples is flow injection analysis (FIA).[14] Its enormous popularity is due to several facts: it is based on sound principles of fluid mechanics and most of the theoretical background of this technique is related to or identical with chemical engineering processes. FIA can be implemented at minimum cost, which means that good results can be obtained very economically. Almost all batch laboratory operations can be performed in the FIA mode, which makes it very acceptable to analytical chemists and process engineers.

The basic idea underlying FIA is simple. If a concentration *step* is introduced into a flowing stream, a sensor placed downstream from the point of injection registers a rising concentration (Figure 1-9) and produces a signal according to its own response characteristics. The property of the concentration step is that, after some time, a steady-state response is reached and the analytical information obtained from the known relationship between the sensor output and the concentration of the injected sample (calibration curve). As long as the sample is *large*, so that it reaches the steady state (depending on the experimental conditions), it does not matter at which *time* the reading is taken. Thus the information acquisition is independent of time and sample size, provided that they are both *large enough* to allow operation in a steady state. There may be some dependence of the sensor output on the flow rate. However, there are sensors (e.g., optical or potentiometric) which are flow-independent. Consequently, in many cases, even the flow rate does not have to be strictly controlled.

Let us now consider the case when a concentration *pulse* is injected. As

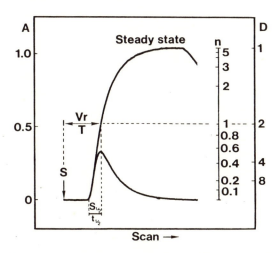

Figure 1-9. Steady-state and transient response in an FIA system. Quantity A is the amplitude of the signal and is proportional to the concentration, V_r is the reactor volume, T the residence time, D the dispersion, $S_{1/2}$ ($t_{1/2}$) the sample volume (time) necessary to reach dilution $D = 2$, and n is the number of volumes $S_{1/2}$ needed to reach the indicated value of D. (Reprinted from Ref. 14, p. 37 with permission of Wiley.)

it passes the sensor a *peak response* is obtained (Figure 1-10). If the sample volume (pulse width) is larger or the flow rate is lower, the peak is shifted toward longer times on the response curve, and *vice versa*. For a given sample volume, readings at different times are taken at concentrations below the peak value. Thus, the information acquisition process for a pulse injection depends also on timing, sample volume, and flow rate. This is apparently a worse situation than that for steady-state measurement. With even moderately priced pumps it is possible to maintain the flow rate accurately (to within 0.5% in a realistic time interval). For a mechanical injector the sample volume is given by the volume of the injection loop, which is constant (but variable on demand), while the distance between the injector and the sensors is also fixed. With these parameters constant the peak time is also constant, but *we may choose* to take the reading at any time before or after the peak time. The advantage is evident: For a non-reacting system the decrease (and increase) in concentration, as registered by the sensor, is the result of a very precisely controlled *dilution* of the sample with the carrier stream. However, if the carrier stream contains a reagent, then the concentration change is due to both *dilution* and *reaction*. Thus by taking readings at different times we are effectively sampling a series of reaction mixtures of different dilution and different degree of conversion. Furthermore, on the time scale of the experiment (as given by the flow rate and the distance between the injector and sensor) such a reaction

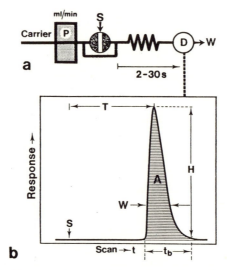

Figure 1-10. A simple FIA manifold (a) and the response (b) to the injection of a concentration pulse. Quantity H is the peak height, W the peak width, T the residence time, and A is the peak area. In the manifold (a) P is a pump, S the injected volume, D the sensor, and W is the waste. (Reprinted from Ref. 14, p. 20 with permission of Wiley.)

may or may not be "instantaneous." Thus the time becomes an additional experimental variable and for "slow" reactions we can even follow the *kinetics*. Therefore, by abandoning the notion of conducting the measurement in a steady state we have gained the following advantages: reduced sample volume (typically 10–100 μl), shorter information acquisition time (normally 15–60 s), and the possibility of performing kinetic experiments. All these facets are very important for sensor applications, because they allow efficient and economical testing of the performance of sensors in addition to the advantages discussed at the beginning of this chapter, namely, calibration and safety. We must also realize that the transient mode of operation can be superimposed on steady-state operation in the FIA mode. Thus, an expensive reagent (such as an enzyme) can be introduced on demand as a broad peak (pseudosteady state) into the main carrier stream and the sample injected as a pulse. Flexibility is one of the strongest points of this technique.

The simplest FIA arrangement is shown in Figure 1-10. In order to perform a chemical reaction, separation, etc., prior to the detection step, a separate reagent line can be added as shown, for example, in Figure 1-11. It should be noted that any number of these peaks can be designated as calibration points. The trade-off is obviously between the time the system spends in the "calibration mode" and in the measuring mode. Almost all

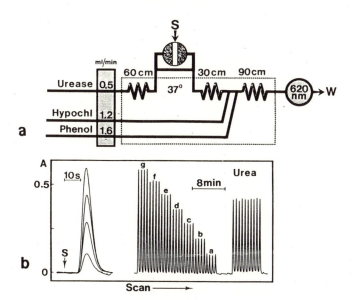

Figure 1-11. Example of the FIA manifold (a) and of the record (b) of urea determination by spectrophotometric detection. The sample volume is 30 μl. (Reprinted from Ref. 14, p. 151 with permission of Wiley.)

operations used in an analytical laboratory have been adapted to FIA and this technique has been used in conjunction with all types of chemical sensors, mostly to the mutual benefit of both.

1.3.2. Other Systems Applications

The different arrangements involving sensors and sample manipulations are too many to be listed and discussed here. A book on the use of electrochemical detectors in flow systems has recently been published.[15] The use of a systems approach can significantly improve detection limits by the *accumulation of the analyte*. This is undertaken in the form of an integration step (meaning physical integration) during which the analyte is preconcentrated. An example is electrochemical stripping analysis used for trace analysis (about 10^{-9} M) of heavy metals. In the preconcentration step (Figure 1-12) the heavy metal(s) is (are) electroplated from the sample

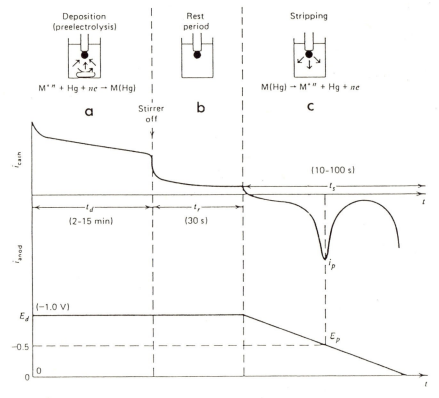

Figure 1-12. Principle of an anodic-stripping voltammetric experiment. (Reprinted from Ref. 16, p. 414 with permission of Wiley.)

into the electrode (typically Hg film) and in the second stage of the analysis it is quantified by measuring the potential or current (or even conceivably optically) corresponding to the anodic dissolution of the amalgam.[16]

Another example involving temporal integration is the assay for trace amounts of toxic substances (in gaseous or liquid samples) based on monitoring the oxygen uptake of a bacterial culture in a reactor.[17] In the absence of toxic compounds the bacterium has a certain consumption of oxygen and the partial pressure of oxygen as monitored by an oxygen sensor is constant. As soon as the toxic compound is introduced the oxygen consumption increases, the result of the bacterium dying off. A system of this type has even been extended to the assay of mutagens.[18]

Another example of the increased sensitivity of a systems approach is the "enzymatic thermistor." The enzymatic reaction is performed in a flow-through microreactor, which enables a sufficiently large amount of heat to be produced and then measured downstream with a thermistor. In this case the much higher sensitivity (and lower detection limit) has been bought for the price of the added complexity of the flow-through reactor.

It is possible to view as sensor systems many analytical operations which are not normally associated with the notion of sensing. For example, the automatic on-line use of a mass spectrometer could be described as a sensor system with the mass spectrometer being a sensor. This may appear to be a really far-fetched example of a sensor system, so we shall examine a gas chromatograph with, for instance, a flame ionization detector (FID). It is possible to describe this situation as a sensor system with FID being an almost nonselective sensor and all the selectivity residing at the column. Again chromatographers would probably object to this classification, so let us consider a spectrophotometer in which the chemical reaction in a cuvette gives rise to a change of optical absorbance. Again we would describe this operation as a conventional spectrophotometric measurement until we alter the hardware in such a way that the light is removed from the spectrophotometer via an optical fiber and the measurement conducted outside the spectrophotometer. Now the otherwise identical operation will be called optical sensing.

It will be seen later that some sensors are difficult or impossible to operate in a continuous mode but can function well in a discrete mode. One example comprises sensors using immunochemical selectivity in which the binding is virtually irreversible under the sensing conditions. Such sensors can be recovered by suitable postsensing treatment.

Miniature valves, pumps, flow chambers, conduits, and so on can be fabricated using technology at the micron level, hence sensor systems provide an excellent alternative to direct sensing. Of course, all operations performed within one data-point cycle, as well as the continuous management of the sensor system, are invariably subject to microprocessor control.

It can be safely stated that almost every type of chemical sensor discussed in this book can be used in a sensor system configuration and its performance can hence be significantly improved.

There is, of course, a price to be paid for all these advantages. The most important one, the frequency-of-sampling limitation, has already been mentioned. Sensor systems also tend to be bulkier than simple sensors due to the added hardware. Furthermore, because it is necessary to withdraw sample aliquots which cannot usually be returned, the total consumption of the sample must be carefully considered, e.g., in medical applications.

Glossary for Chapter 1

a	Activity
C	Concentration
D_s	Diffusion coefficient
E_{out}	Sensor output
F	Faraday constant
f	Activity coefficient
G	Free energy
G°	Standard free energy
K	Equilibrium constant
$K_{x,i}$	Selectivity coefficient
K_m	Michaelis–Menten constant
n_i	Number of moles of species i
R	Gas constant
T	Absolute temperature
β	Adsorption isotherm
ϕ	Thiele modulus
Γ	Surface concentration
μ_i	Chemical potential of species i
$\mathfrak{R}$	Response function

References for Chapter 1

1. K. E. van Holde, *Physical Biochemistry*, 2nd ed., Prentice-Hall, New York, 1985.
2. H. Stieve, *Sensors and Actuators* 4 (1983) 689.
3. D. R. Absolom and C. J. van Oss, *CRC Crit. Rev. Immunol.* 6 (1986) 1.
4. M. A. Sharaf, D. L. Illman, and B. Kowalski, *Chemometrics*, Wiley, New York, 1986.
5. S. D. Caras, J. Janata, D. Saupe, and K. Schmidt, *Anal. Chem.* 57 (1985) 1917.
6. L. E. Hood, I. L. Weissman, W. B. Wood, and J. H. Wilson, *Immunology*, 2nd ed., Benjamin & Cummings, Menlo Park, 1984.
7. B. L. Liu and J. S. Schultz, *IEEE Trans. Biomed. Eng.* BME-33 (1986) 133.

8. Various authors, Immobilization of proteins, in: *Methods in Enzymology* (K. Mosbach and B. Danielsson, eds.), Vols. 135 and 136, Academic Press, Orlando, Florida, 1987.

9. M. A. Arnold, *Ion-Sel. Electrode Rev.* 8 (1986) 85.

10. S. D. Caras, D. Petelenz, and J. Janata, *Anal. Chem.* 57 (1985) 1920.

11. Y. Hanazato, M. Nakako, M. Maeda, and S. Shiono, *Anal. Chim. Acta* 193 (1987) 87.

12. Y. Miyahara, T. Moriizumi, and K. Ichimura, *Sensors and Actuators* 7 (1985) 1.

13. S. J. Pace, *Sensors and Actuators* 1 (1981) 499.

14. J. Ruzicka and E. H. Hansen, *Flow Injection Analysis*, Wiley, New York, 1988.

15. K. Stulik and V. Pacakova, *Electroanalytical Measurements in Flowing Liquids*, Wiley, New York, 1987.

16. A. J. Bard and L. Faulkner, *Electrochemical Methods*, Wiley, New York, 1980.

17. K. Matsumoto, H. Seijo, T. Watanabe, I. Karube, I. Satoh, and S. Suzuki, *Anal. Chim. Acta* 105 (1979) 429.

18. I. Karube and M. Suzuki, *Biosensors* 2 (1986) 343.

Thermal Sensors

The first law of thermodynamics tells us that any process in which the internal energy of the system changes is accompanied by absorption or evolution of heat. The class of chemical sensors to be discussed in this chapter uses the heat generated by a specific reaction as the source of analytical information. The important point to realize is that these sensors represent a form of *in situ* microcalorimetry which could be otherwise performed in a batch mode. In the case of enzyme thermistors the equivalent batch technique would be called adiabatic calorimetry and, in the case of pyroelectric sensors, it would be a heat-flow isothermal calorimetry.

There are two properties of heat which are quite unique with respect to any other physical parameter: heat is totally nonspecific and cannot be contained, i.e., it flows spontaneously from the warmer (T_1) to the colder (T_2) part of the system. From the sensing point of view this defines the task of designing a thermal sensor in a special way. The general strategy is to place the chemically selective layer on top of a thermal probe and measure the heat evolved in the specific chemical reaction taking place in that layer, either as the change in temperature of the sensing element or as the heat flux through the sensing element (Figure 2-1). The heat is evolved continuously, so thermal sensors are in a *nonequilibrium state* and their signal is obtained from a *steady-state* situation.

At constant pressure the generated heat Q_p is equal to the enthalpy change ΔH [see equation (A-7)], which is related to the change in temperature. In a closed system [adiabatic change; see equation (A-9)]

$$dT = -dH/C_p \qquad (2\text{-}1)$$

where C_p is the heat capacity. If C_p is constant over a given temperature range, the enthalpy change can be calculated from such a measurement for any temperature within that range.

A schematic diagram of a thermal chemical sensor is shown in Figure 2-1. The sensing element can be any temperature probe. Owing to their sensitivity thermistors are preferred to thermocouples. There are some conflicting requirements built into the operation of these sensors: in order for the sensor to interact with the chemical species it must exchange matter, i.e., it must be a thermodynamically open system. On the other hand, in order to obtain a maximum response it should be as adiabatic (thermally insulated) as possible. This conflict presents some difficulties in the design of the optimum configuration.

In order to provide a steady-state signal the heat must be evolved continuously. An ideal reaction for this is an enzymatic reaction, which combines substrate specificity with high amplification factor. Thus, an enzyme-containing layer is deposited over the thermal probe and the substrate allowed to diffuse into this from the sample solution. As it diffuses it reacts according to a general Michaelis–Menten equation [see equation (1-37)] and the molar heat equal to the enthalpy of that reaction is evolved:

$$E + S \leftrightarrow ES \rightarrow P \qquad (-\Delta H)$$

Molar heats of some enzymatic reactions are given in Table 2-1. In order to relate the output of the thermal sensor to the bulk concentration of the substrate which is being enzymatically converted, equation (2-1) would have to be incorporated into the diffusion-reaction mechanism discussed in Section 1.2.2. This would be exceedingly difficult. The reason is that, unlike *mass flow*, *heat flow* through the sensor is anisotropic, which means that the above equations would have to be solved simultaneously with the three-dimensional heat flow equation

$$0 = \partial T/\partial t = \Delta H[\Re_x \partial^2/\partial x^2 + \Re_y \partial^2/\partial y^2 + \Re_z \partial^2/\partial z^2]T \qquad (2-2)$$

where $\Re = D/MK_p$.

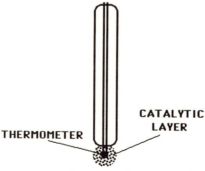

CATALYTIC LAYER

THERMOMETER

Figure 2-1. Schematic diagram of a thermal catalytic sensor. The element T can be any temperature-measuring device.

Table 2-1. Molar Enthalpies of Some Enzyme-Catalyzed Reactions[a]

Enzyme	Substrate	$-\Delta H$ (kJ mol^{-1})
Catalase	Hydrogen peroxide	100.4
Cholesterol oxidase	Cholesterol	52.9
Glucose oxidase	Glucose	80
Hexokinase	Glucose	27.6
Lactate dehydrogenase	Pyruvate	62.1
Trypsin	Benzoyl-L-arginineamide	27.8
Urease	Urea	6.6
Uricase	Uric acid	49.1

[a] Reprinted from Ref. 2 with permission from the American Chemical Society.

Such an equation has not yet been solved. A further complicating factor is that every reaction (including the protonation equilibria) contributes to the overall thermal effect. This would make even numerical solutions too complicated. The corollary of the anisotropic heat flow is that the overall temperature change in the thermal sensing element is relatively small (typically 0.1–1 m°C) because only a fraction of the total heat contributes to the useful temperature change. There are two thermal probes which we shall consider for monitoring thermal processes: thermistors and pyro-electric devices. Although thermopiles (thermocouples) have been proposed as possible thermal transducers, their sensitivity is much lower (about 10 μV per °C).

2.1. Thermistors

Thermistors are the probes of choice because they are stable (± 0.05 °C/year), inexpensive, sensitive, chemically inert, and small.[1] They are made of high-temperature sintered oxides (such as BaO/SrO) covered with a glass coating (Figure 2-2). They cover a temperature range from -80 °C to $+350$ °C and have negative (NTC) or positive (PTC) temperature coefficient. Typical resistances range from 100 Ω to MΩ. Their response is related to the bandgap energy E_g of the oxide semiconductor:

$$R_T = R_c \exp(\pm E_g/2kT) \tag{2-3}$$

This latter equation can be differentiated with respect to temperature to yield

$$dR_T/dT = \pm \frac{R_c E_g}{2kT^2} \exp(\pm E_g/2kT) \tag{2-4}$$

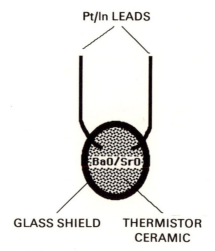

Pt/In LEADS

BaO/SrO

GLASS SHIELD   **THERMISTOR CERAMIC**

Figure 2-2. Diagram of a shielded bead thermistor.

The relative resistance change (thermistor coefficient α) is obtained by dividing equation (2-4) by equation (2-3):

$$\frac{1}{R_T}\frac{dR_T}{dT} = \alpha = \pm\frac{E_g}{2kT^2} \tag{2-5}$$

A typical value of α is $\pm 4\%$ per °C. The resistance change with temperature is shown in Figure 2-3. Owing to their nonlinear response, thermistors are usually linearized by placing a resistor of similar nominal value in parallel with the thermistor, with resulting loss of sensitivity

$$\frac{1}{R_T + R_p}\frac{dR_T}{dT} = \alpha \tag{2-6}$$

This is not generally necessary for thermochemical sensor applications, because the temperature range involved is small. A sensitivity of $10^{-4}\ \Omega/°C$ can be achieved with a conventional Wheatstone bridge. Thermistors are often used in matched pairs, in which case the thermal common-mode rejection ratio ($CMRR_T$) is defined as

$$(CMRR_T) = \frac{T_{final} - T_{initial}}{\Delta T_{apparent}} \simeq \frac{\langle\alpha\rangle}{\Delta\alpha} \tag{2-7}$$

Typical values of $CMRR_T$ are 10–100.

The design of enzyme thermistors is similar to that of enzyme electrodes,

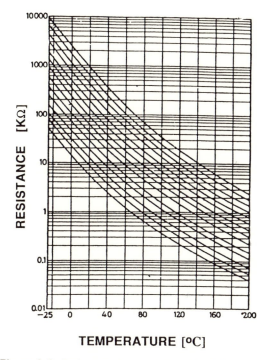

Figure 2-3. Resistance characteristics of bead thermistors.

and identical enzyme immobilization procedures are used. An example of this type of device is the enzyme thermistor using immobilized glucose oxidase (GOD).[2] The reactions are

$$\beta\text{-D-glucose} + H_2O + O_2 \xrightarrow{\text{GOD}} H_2O_2 + \text{D-gluconic acid} \qquad \Delta H_1$$

and

$$H_2O_2 \xrightarrow{\text{catalase}} \tfrac{1}{2}O_2 + H_2O \qquad \Delta H_2$$

The total enthalpy change is the sum of the two partial enthalpies ΔH_1 and ΔH_2, approximately -80 kJ mol^{-1}. The enzymes are immobilized on the tip of a thermistor, which is then partially enclosed in a glass jacket in order to reduce heat loss to the surrounding solution (Figure 2-4). A second thermistor with similarly immobilized bovine serum albumin is used as a reference device in the opposing arm of the Wheatstone bridge. The response of this arrangement is shown in Figure 2-5. The detection limit is modest, about 3 mM. Similar devices with similar performance characteristics have been constructed using urease and other enzymes.

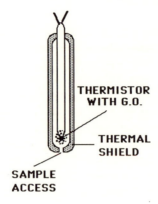

Figure 2-4. Schematic diagram of a glucose thermistor. (According to Ref. 2.)

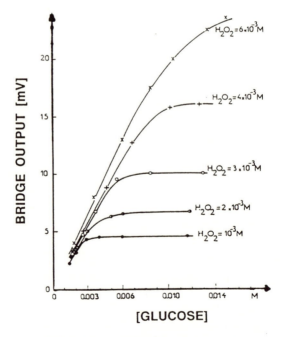

Figure 2-5. Response of the glucose oxidase thermistor to glucose in the presence of different hydrogen peroxide concentrations. (Reprinted from Ref. 2 with permission of the American Chemical Society.)

2.2. *Pyroelectric Devices*[3,4]

Piezo- and pyroelectricity have their origin in the material properties of certain crystals. The two effects are similar, so they will be discussed together in this chapter. The pyroelectric effect is used in infrared detectors and pyrometry. Their use in chemical sensing will be examined in Chapters 3 and 4.

Piezoelectricity couples electromechanical effects. A good analog model is a two-dimensional hexagonal "crystal" consisting of electric charges $q^\pm$ (ions) connected by mechanical springs of length l (Figure 2-6). It is assumed that the dipoles **AB**, **CD**, and **EF** are not mutually interacting. We note that such a crystal does not possess central symmetry, i.e., the projection of any ion through the center O (e.g., $\mathbf{B}^-$ to $\mathbf{F}^+$) is to the

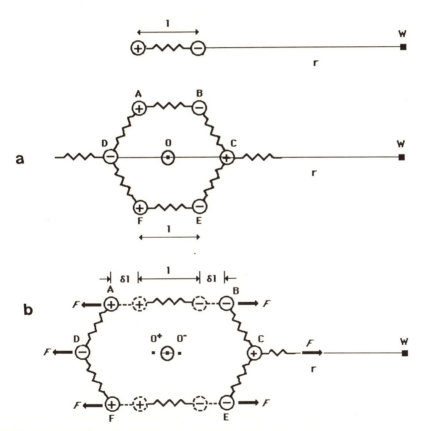

Figure 2-6. Model of a hypothetical ionic solid. The elastic forces between ions are represented by coils of length l. (a) Electric field at point W situated distance r from the elementary crystal. (b) The same solid subjected to stress F. (According to Ref. 4.)

ion of *opposite* polarity. There are 21 crystallographic systems which do not possess a center of symmetry; 20 exhibit piezoelectricity. PXE ceramics (such as $PbTiO_3$ and $PbZrO_3$) and some stressed polymers (such as PVF_2) are also excellent piezoelectrics. The piezoelectric effect has its origin in the microcrystalline regions in these materials in which the permanent electric dipole has been "frozen" along the poling axis during the fabrication, by applying a strong electric field while subjecting the material to mechanical stress. They can be fabricated in different shapes and sizes and exhibit high chemical resistance.

The electric field E due to an electric dipole at any point on the axis of the dipole at distance $r \gg l$ from the dipole is given by

$$E = \chi/2\pi\varepsilon_0 r^3 \tag{2-8}$$

where χ is the electric dipole moment equal to ql. The reference point W is chosen at distance r from the two-dimensional "crystal." The electric field at this point is the sum of the contributions of the dipolar fields from **AB**, **CD**, and **EF**:

$$E = \frac{(ql)_{AB} - (q2l)_{CD} + (ql)_{EF}}{2\pi\varepsilon_0 r^3} = 0 \tag{2-9}$$

When a symmetrical force acting along the axis **OW** is applied, the "crystal" will be deformed and, to a first approximation, each ion will be displaced by an equal distance δl. Addition of all field contributions from the individual ions, while keeping track of the charges yields, for the total field at point W,

$$E = 2q \, \delta l/2\pi\varepsilon_0 r^3 \tag{2-10}$$

The dipole moment **p** created by the displacement of charge due to the applied force is $\chi = 2q \, \delta l = 2ql\psi$, where the longitudinal strain ψ equals the relative longitudinal elongation $\delta l/l$.

For a real crystal consisting of N atoms per unit volume, the electric polarization is the sum of the dipole·moment contributions from all displaced atoms, namely,

$$\chi_E = qN = 2qlN\psi \tag{2-11}$$

The ionic charge q, atomic distance l, and number of atoms are characteristic properties of the crystal, so they are incorporated together in the quantity $\sigma = 2qlN$, the stress tensor.

By the same argument it can be easily verified that the "crystal" having center of symmetry does not produce an electric field when it is distorted mechanically. It is important to note that if the ionic charges differ, then the residual field (polarization) exists within the "crystal" even without mechanical distortion. The presence of permanent polarization in addition to the lack of central symmetry is a prerequisite of the pyroelectric effect. Thus a typical piezoelectric material, quartz, does not exhibit pyro-electricity. The pyroelectric material of choice is $LiTaO_3$. Both effects are highly anisotropic, therefore great care must be taken in the design of proper mounting for the crystals.

The total free energy G of a crystal (system) can be expressed in terms of the first and second laws of thermodynamics [see equations (A-6) and (A-13)]:

$$G = U + PV - TS \qquad (2\text{-}12)$$

or as the total differential

$$dG = dU + P\,dV + V\,dP - T\,dS - S\,dT \qquad (2\text{-}13)$$

For a reversible system [see equations (A-2) and (A-10)]

$$dU = T\,dS - dW \qquad (2\text{-}14)$$

The work done by the piezoelectric crystal $(-dW)$ is at the expense of mechanical $(\sigma\psi)$ and electric (ED) polarization:

$$\Delta W = \sigma\psi + ED \qquad (2\text{-}15)$$

where D is the electric displacement. If equation (2-15) is differentiated and combined with equations (2-13) and (2-14), we obtain

$$dG = -S\,dT - \sigma\,d\psi - \psi\,d\sigma - E\,dD - D\,dE + P\,dV + V\,dP \qquad (2\text{-}16)$$

The pressure applied to the crystal at constant volume $(P\,dV = 0)$ causes changes in the mechanical strain $d\psi$ and in the displacement of charge dD,

$$V\,dP = \sigma\,d\psi + E\,dD \qquad (2\text{-}17)$$

Equations (2-16) and (2-17) yield

$$dG = -S\,dT - \psi\,d\sigma - D\,dE \qquad (2\text{-}18)$$

The Maxwell relationships are then obtained by partial differentiation of equation (2-18):

$$\left(\frac{\partial G}{\partial T}\right)_{\sigma, E} = -S, \qquad \left(\frac{\partial G}{\partial \sigma}\right)_{T, E} = -\psi, \qquad \left(\frac{\partial G}{\partial E}\right)_{T, \sigma} = -\mathbf{D} \qquad (2\text{-}19)$$

The second partial differentiation yields the expression for the pyroelectric vector **p**:

$$\frac{\partial S}{\partial E} = -\frac{\partial^2 G}{\partial T \partial E} = \frac{\partial D}{\partial T} = \mathbf{p} \qquad (2\text{-}20)$$

i.e., electric displacement with temperature. Similarly, the piezoelectric coefficient **d** is given by

$$\frac{\partial D}{\partial \sigma} = -\frac{\partial^2 G}{\partial E \partial \sigma} = \frac{\partial \psi}{\partial E} = \mathbf{d} \qquad (2\text{-}21)$$

which is the mechanical strain caused by the electric field. The mechanical displacement caused by temperature is the thermal expansion coefficient α_T, given by

$$\frac{\partial S}{\partial \sigma} = -\frac{\partial^2 G}{\partial T \partial \sigma} = \frac{\partial \psi}{\partial T} = \alpha_T \qquad (2\text{-}22)$$

From equations (2-20) and (2-21) we see that the electric displacement can be caused by the change of temperature $(\partial D/\partial T)$ (pyroelectricity) or by applying the mechanical stress $(\partial D/\partial \sigma)$ to the crystal, i.e., the piezoelectricity. It can also be caused by applying an external electric field, for example, by placing the crystal between the plates of a capacitor. Thus

$$D = F(E,\ T,\ \text{and } \sigma) \qquad (2\text{-}23)$$

Differentiation of equation (2-23) yields

$$dD = \left(\frac{\partial D}{\partial E}\right)_{T,\sigma} dE + \left(\frac{\partial D}{\partial T}\right)_{\sigma,E} dT + \left(\frac{\partial D}{\partial \sigma}\right)_{E,T} d\sigma \qquad (2\text{-}24)$$

which can then be expressed in the form

$$dD = \varepsilon\, dE + \mathbf{p}\, dT + \mathbf{d}\, d\sigma \qquad (2\text{-}25)$$

This quantifies the electric displacement of the material in terms of its dielectric constant, and its pyroelectric and piezoelectric coefficients.

Pyroelectricity depends on the crystal orientation. If the external electric field is applied along the pyroelectric axis, then the dielectric constant in this direction $\varepsilon_\mathbf{p}$ is [from equation (2-24)]

$$(\partial D/\partial E)_{T,\sigma} = \varepsilon_\mathbf{p} \qquad (2\text{-}26)$$

For pyroelectric devices (without stress), the third term in equation (2-25) vanishes and the average volume electric displacement $\langle dD \rangle$ becomes

$$\langle dD \rangle = \mathbf{p}\, dT + \varepsilon_{\mathbf{p}}\varepsilon_0\, dE \tag{2-27}$$

where ε_0 is the dielectric constant of a vacuum. Thus, the pyroelectric effect causes displacement of charge (generation of potential difference $V_0 - V$). The crystal forms a pyroelectric capacitor C_p of area A and thickness d:

$$C_p = A(\varepsilon_{\mathbf{p}}\varepsilon_{\mathbf{p}})/d \tag{2-28}$$

On the assumption that $\mathbf{p}$ is independent of temperature in the interval from T to T_0, and because for a parallel-plate capacitor

$$dE = (V - V_0)/d \tag{2-29}$$

the total charge ΔQ is obtained from equation (2-27),

$$Q = -C_p(V - V_0) + \mathbf{p}(T - T_0) \tag{2-30}$$

The time differential of equation (2-30) yields the pyroelectric current i_p in the form

$$i_p = dQ/dt = -C_p\, dV/dt + \mathbf{p}\, dT/dt \tag{2-31}$$

Subject to a steady state $dV/dt = 0$, in which case

$$i_p = \mathbf{p}\, dT/dt \tag{2-32}$$

If the heat flux exists and is governed by the equation

$$dT/dt = -\Delta H/C_p M \tag{2-33}$$

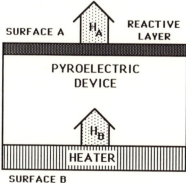

Figure 2-7. Heat flow in a pyroelectric capacitor. (According to Ref. 3.)

then the pyroelectric voltage V_p is generated along the internal resistance of the crystal R_p and is given by

$$V_p = R_p i_p = -\mathbf{p} R_p \,\Delta H / C_p M \qquad (2\text{-}34)$$

where C_p is the heat capacity of the crystal, M its thermal mass, and ΔH is the enthalpy of the reaction.

The equivalent circuit for a pyroelectric crystal is shown in Figure 2-7. The fundamental difference between the pyroelectric and thermoresistive methods of measurement is that the former measures the difference in *heat flux* while the latter measures the temperature difference. This results in a three-order-of-magnitude increased sensitivity of pyroelectric devices over thermistors.

In contrast to piezoelectric devices, pyroelectric devices are operated in a DC mode. Infrared detectors using pyroelectric crystals often use a sinusoidally varying heat flux produced by chopping the measured radiation in order to increase the signal-to-noise ratio. The advantage of this mode of operation and its implementation in chemical sensors is not obvious at present.

Some practical devices are now presented. A Z-cut crystal of lithium tantalate ($LiTaO_3$), $230\ \mu m$ thick with NiCr contacts (Figure 2-8), has been used as a pyroelectric device. The chemically selective material which generates the heat is applied to one of the electrodes. So far there have been very few applications of pyroelectric chemical sensors. A study exists in which a hyperbolic temperature program was applied to the crystal with preadsorbed gas and the heat of desorption measured. The sensitivity of this measurement was approximately $10\ \mu W$, which means that direct

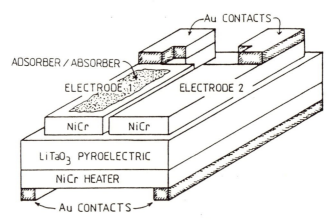

Figure 2-8. Schematic of a dual pyroelectric sensor. (Reprinted from Ref. 3 with permission of Academic Press.)

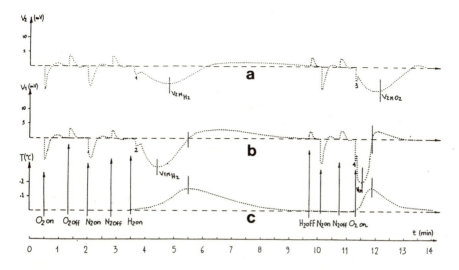

Figure 2-9. Response of the pyroelectric gas sensor. The lower trace (a) shows the temperature program. Curves (b) and (c) are from the Pd and Au coatings, respectively. (Reprinted from Ref. 3 with permission of Academic Press.)

measurement of desorption heat of less than a monolayer of gas could be measured with a signal-to-noise ratio of 10. The example of the response is shown in Figure 2-9. The use of an enzyme layer as the selective material/ heat generator is a real possibility but, so far, such a device has not been reported.

2.3. Catalytic Gas Sensors

These devices are conceptually similar to enzymatic transistors. Heat is liberated as a result of catalytic reaction taking place at the surface of the sensor and the related temperature change inside the device is measured. On the other hand, the chemistry involved is very similar to that of high-temperature conductimetric oxide sensors (Section 4.3.3). Catalytic gas sensors have been designed specifically for the detection of subthreshold concentrations of flammable gases in ambient air with the safety of mining operations in mind. Although the name has been reserved for sensors using porous catalytic layer, they are often called *pellistors* as a group.[5]

As far as the structure is concerned, they are one of the simplest sensors available (Figure 2-10). The platinum coil is imbedded in a pellet of ThO_2/Al_2O_3 coated with porous catalytic metal, palladium, or platinum. The coil acts both as a heater and as a resistance thermometer. It is

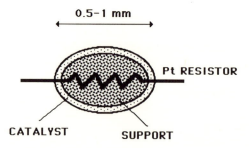

Figure 2-10. Cross section of a catalytic gas sensor: a pellistor.

possible to use any type of temperature sensor without changing the basic concept of this sensor.

When the combustible gas reacts at the catalytic surface, heat evolved from this reaction increases the temperature of the pellet and of the platinum coil, thus increasing its resistance. The pellistor is usually connected in one arm of the Wheatstone bridge, which provides the output signal. For small temperature changes, the out-of-balance signal of the bridge is then proportional to the rate of the raction and to the reaction heat ΔH:

$$V = K(d[S]/dt)\,\Delta H \qquad\qquad (2\text{-}35)$$

The proportionality constant K includes both the heat capacity of the pellistor and the factor related to the convective heat loss. There are two principal modes of operation: isothermal and nonisothermal. In the isothermal mode the out-of-balance bridge signal is used in a negative feedback, which reduces the current (i.e., the power input) into the Pt wire. Thus the increment of reaction heat supplied to the pellet is compensated by the equal reduction in the resistive power ΔW given by

$$\Delta W = -(d[G]/dt)\,\Delta H \qquad\qquad (2\text{-}36)$$

The nonisothermal mode can be used in conjunction with a diffusion barrier. For an arrangement where the supply of the gas G to the surface of the pellet is controlled by diffusion, the bridge output is proportional to the gas concentration,

$$V = KD_T\,\Delta H[G] \qquad\qquad (2\text{-}37)$$

The response is usually expressed on the scale of percent of lower explosion limit (%LEL) which, for methane, is %100 LEL $= 5\%$ v/v in air. On the LEL scale the dynamic range is 10–100%. The fundamental problem, also common to other types of catalytic sensor, is catalyst poisoning by, for instance, organosulfur compounds. The operating temperature is typically 200–300 °C with a time response of around 1 s.

Glossary for Chapter 2

C_p	Heat capacity
D	Electric displacement
D_T	Diffusion coefficient of heat
d	Piezoelectric coefficient
E	Electric field
E_g	Bandgap energy
H	Enthalpy
i_p	Pyroelectric current
k	Boltzmann constant
M	Thermal mass
p	Pyroelectric vector
q	Charge
$R_{T,c}$	Resistance
α	Thermistor coefficient
α_T	Thermal expansion coefficient
χ	Dipole moment
ψ	Strain
σ	Stress

References for Chapter 2

1. R. S. C. Cobbold, *Transducers for Biomedical Instruments*, Wiley, New York, 1974.
2. Can Tran-Minh and D. Vallin, *Anal. Chem.* 50 (1978) 1874.
3. J. N. Zemel, in: *Solid State Chemical Sensors* (J. Janata and R. J. Huber, eds.), Academic Press, New York, 1985.
4. V. M. Ristic, *Principles of Acoustic Devices*, Wiley, New York, 1983.
5. S. J. Gentry, Catalytic devices, in: *Chemical Sensors* (T. E. Edmonds, ed.), Chapman and Hall, New York, 1988.

3

Mass Sensors

Change of mass can be viewed as a general feature of the interaction of the chemical species with the sensor. From the measurement point of view the determination of mass is generally called gravimetry. Although scales and balances are standard equipment in any laboratory, they are not usually regarded as sensors. On the other hand, when we talk about microbalances and microgravimetry[1] we regard them as sensors. Owing to their small size, high sensitivity, and stability, piezoelectric crystals have been used as microbalances, namely, in the determination of thin-layer thickness and in general gas-sorption studies.[2] The incorporation of various chemically sensitive layers has enabled the transition from microbalance to mass sensor and has resulted in the explosive growth of piezoelectric sensors in recent years.[3] The major advantages of mass sensors are their simplicity of construction and operation, their light weight, and the low power required. Measurement of the frequency shift is one of the most accurate types of physical measurement. Compared with electrochemical sensors the measurement is conducted in a monopolar mode, i.e., only a single physical probe is necessary. Mass sensors have high sensitivity and can be used for a very broad range of compounds. However, the corollary is high vulnerability to interferences. Piezoelectric crystals are relatively inexpensive and readily available.

The advantage of surface acoustic wave devices is great design flexibility and the fact that the sensing part of the device, the waveguide, is separate from the piezoelectric transducer/receiver. On the other hand the physics is complex, which makes it difficult to obtain a general explicit relationship between the added mass and the change in the output.

The optimum range of applicability of this type of sensor can be deduced from some general aspects of the behavior of these devices and from the interaction between the basic species and the sensor,

$$S + X \Leftrightarrow SX$$

Clearly, the signal will be obtained only if the above interaction results in a net change of mass of the chemically selective layer attached to the crystal. Thus, an equilibrium binding represented by the above reaction will yield a measurable signal. On the other hand, if the interaction is a displacement of one species with another, i.e., exchange or catalytic reaction, the sensor surface is only a temporary host to the interacting species and the net changes of mass may be very small.

Another limitation arises from the operation of these devices in a liquid phase, to be discussed later in more detail. In such media the output signal of the sensor is affected not only by the incremental mass due to the above reaction, but also by frictional effects at the sensor–liquid interface that affect the energy loss. Water is the required environment for most biological reactions, so it follows that the biological means of achieving chemical selectivity are not generally suitable for this type of sensor.

3.1. Piezoelectric Sensors

There are several types of material that exhibit the piezoelectric effect. The crystallographic requirements for this effect to exist have been discussed in Section 2.2. α-Quartz is the material selected for most piezoelectric sensor applications, because it is inexpensive and has a relatively high piezoelectric coefficient; it also possesses a hexagonal crystallographic structure with no center of symmetry (Figure 3-1). It follows from equations (2-21)–(2-25) that the magnitude of the piezoelectric coefficient and its temperature dependence depend on the orientation of the cut of the crystal (Figure 3-1) with respect to the main axes. The temperature dependence of the piezoelectric coefficient as a function of the angle of the cut is shown in Figure 3-2. For practical reasons the optimum orientation is chosen so that the crystal exhibits minimum temperature dependence within the operating range of temperatures. For sensor applications under ambient conditions the AT cut (35°15′ inclination in the y–z plane) is the most usual.

Owing to the fact that charge leaks, application of a steady force produces only a transient voltage which can be as high as 20,000 V. For this reason piezoelectricity is almost exclusively studied and used in the AC mode. This fact is, however, very beneficial from the measuring point of view, because the frequency is one of the most precisely measurable quan-

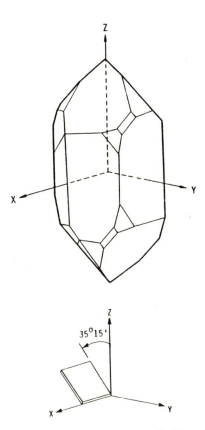

Figure 3-1. The assignment of axes to a quartz crystal with a 35°15′ AT cut. (Reprinted from Ref. 7 with permission of Elsevier.)

tities with precision about $1:10^{10}$ and stability retention over 1 month. The quantitative relationship governing the operation of these sensors therefore links the mass with the change of frequency.

For design purposes it is convenient to represent the piezoelectric microbalance by the equivalent electrical circuit (Figure 3-3). Unfortunately, this concept does not lead to the quantitative relationship between the mass and frequency.

In analogy with *mechanical elasticity* we define motional capacitance and describe it by capacitance C. *Vibrating mass* is identified with motional inductance and represented by electrical inductance L. The frictional energy loss is identified with electrical resistance R. If the electrical admittance is plotted as a function of frequency, a curve is obtained (Figure 3-3) in which the resonance (loss-free oscillation) occurs at frequencies corresponding to the maximum (series) and minimum (parallel) at which the

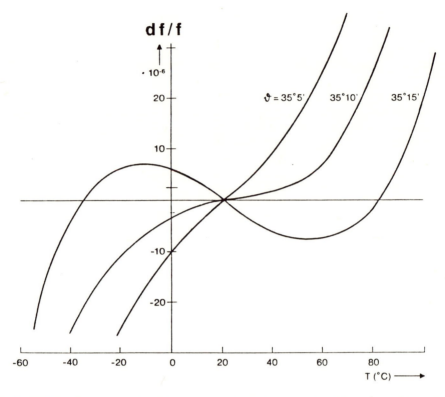

Figure 3-2. Temperature dependence of the relative frequency for different cuts of quartz crystal. (Reprinted from Ref. 7 with permission of Elsevier.)

complex impedance is purely resistive. In the series mode the vibration is perpendicular to the electric field vector:

$$f_s = (1/2\pi)(1/LC)^{1/2} \tag{3-1}$$

while in the parallel mode the vibration is parallel to the electric field in the crystal:

$$f_p = (1/2\pi)[(1/LC) + (1/LC_0) + (R_1 + R_L)^2/L^2]^{1/2} \tag{3-2}$$

where C_0 is the electrode and stray capacitance, which is unimportant in the usual mode of operation (series). The effect of mass on frequency is seen as the effect of increased values of L, mechanical elasticity C, and acoustic load R_L, all of which result in a decrease of the resonance frequency. However, it is difficult to establish explicitly the relationship

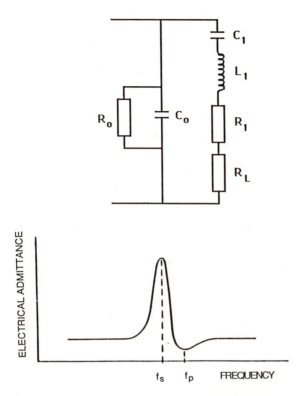

Figure 3-3. Equivalent electrical circuit and resonant frequency of a mounted piezoelectric crystal oscillator: C_0 is the capacitance of the mounted crystal, C_1 and L_1 represent the rigidity and mass of the added material, R_0 is the dielectric, R_1 the mechanical loss, and R_L represents the acoustic (mechanical) load. On the plot, f_s and f_p are the series [equation (3-1)] and parallel [equation (3-2)] resonant frequencies.

between the added mass and the frequency shift. Thus, the equivalent circuits are useful mainly for design purposes, but they do not provide a route for computing the absolute mass.

3.1.1. Sauerbrey Equation

A piezoelectric crystal vibrating in its resonance mode is a harmonic oscillator. For the microgravimetric application it is necessary to develop a quantitative relationship between the relative shift of the resonant frequency and the added mass. In the following derivation the added mass is treated as an "added thickness" of the oscillator, an approach which makes the derivation more intuitively accessible.

Figure 3-4 shows the shear-mode vibration of a quartz crystal of mass M_q and thickness t_q. At resonance the wavelength is

$$\lambda = 2t_q \qquad (3\text{-}3)$$

The shear velocity v_q is defined as

$$v_q = \lambda_q f_q \qquad (3\text{-}4)$$

therefore

$$f_q t_q = v_q/2 \qquad (3\text{-}5)$$

The resonance frequency shift caused by an infinitesimally small change of crystal thickness dt_q is then obtained by differentiating equation (3-5) to obtain

$$df_q = -v_q\, dt/2t_q^2 \qquad (3\text{-}6)$$

Division of equation (3-6) by equation (3-5) yields the relative change of frequency in the form

$$df_q/f_q = -dt_q/t_q \qquad (3\text{-}7)$$

Thus, the relative increase in crystal thickness (Figure 3-4b) lowers the resonant frequency. Equation (3-7) can be expressed in terms of the total mass M and its change dM_q,

$$df_q/f_q = -dM_q/M_q \qquad (3\text{-}8)$$

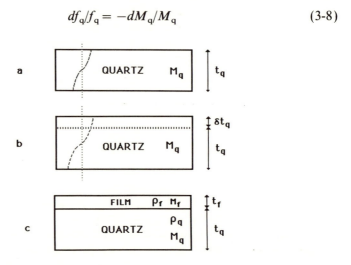

Figure 3-4. Shear wave in (a) a quartz crystal, (b) with incremental increase in crystal thickness, and (c) with added film.

In this derivation the *increment of the foreign mass dM* (uniformly distributed over the crystal surface to form a thin film M_f) is treated as the equivalent change of mass of the crystal. Therefore the approximation $M_q \sim M_f$ is introduced,

$$df_q/f_q = -dM/M_q \qquad (3-9)$$

The relative change of resonant frequency upon deposition of film M_f is then

$$(f_c - f_q)/f_q = -M_f/M_q \qquad (3-10)$$

where f_c is the resonant frequency of the crystal with deposited material. The same equation expressed "per unit area" ($m_f = M_f/\text{area}$) is

$$(f_c - f_q)/f_q = -m_f/m_q \qquad (3-11)$$

It will now be assumed that both the film (or the deposit) and the crystal have uniform density ρ_f and ρ_q, respectively. The mass is the product of thickness and density, so equations (3-5) and (3-11) yield

$$m_f = -(f_c - f_q)\, \rho_q v_q/2f_q^2 \qquad (3-12)$$

If the film density is known, the thickness can be calculated. This is the principle of the "thickness monitor" as used in monitoring film thickness during vacuum deposition or electroplating.

It is further assumed that the shear velocity is identical in the film and in the crystal. This is a relatively crude assumption. It is convenient to define the crystal calibration constant C_f from equation (3-12) as

$$C_f = 2f_q^2/(\rho_q v_q) \qquad (3-13)$$

through which the frequency shift is related to the added mass m_f,

$$\partial f = -C_f m_f \qquad (3-14)$$

Equations (3-12)–(3-14) are different forms of the Sauerbrey equation, which is the fundamental relationship governing piezoelectric sensors. Due to the assumptions made throughout this derivation, the Sauerbrey equation is only semiquantitative. Other factors such as mechanical clamping, damping in the electrical circuit, and temperature also affect the absolute accuracy. For this reason it is necessary to employ calibration curves for quantitative work. In spite of these limitations, the quartz microbalance is an extremely sensitive and very stable sensor. Thus, for example, an AT-cut quartz ($\rho = 2650$ kg m^{-3}; $v_q = 3340$ m s^{-1}) with resonant frequency at $f = 5$ MHz possesses crystal constant $C_f = 5.65$ MHz cm^2 kg^{-1}, which means that a 1 Hz shift corresponds to 17.7 ng cm^{-2} added weight. The dynamic range extends up to 20 μg cm^{-2} which, however, exceeds the thickness limit of the Sauerbrey equation.

Chemical selectivity is provided by some chemical layer, which means that conditions for resonance should be examined for a crystal containing such a layer. The simplest case is a thin film applied to one side of the crystal.[3] At the interface between the crystal (subscript q) and film (subscript f) reflection and refraction of the acoustic energy occurs (Figure 3-4c), in analogy with optical reflection and refraction that occurs at the boundary between two materials of different optical densities. The shear velocity in the crystal and film is, respectively,

$$v_q = (\mu_q/\rho_q)^{1/2} \quad \text{and} \quad v_f = (\mu_f/\rho_f)^{1/2} \tag{3-15}$$

where μ_q and μ_f are the shear moduli of the respective layers. If we assume that there is no frictional loss, i.e., that the resonance condition applies, then

$$\tan(\pi f_c/f_q) = -(\rho_f v_f/\rho_q v_q) \tan(\pi f_c/f_f) \tag{3-16}$$

where $f_c = \omega/2\pi$ is the resonant frequency of the crystal with material deposited on it. The terms $\rho_f v_f$ and $\rho_q v_q$ are the acoustic impedances Z_f and Z_q of the film and crystal, respectively. Their ratio $Z = Z_f/Z_q$ is an important parameter with respect to acoustic matching of the two materials. For optimum resonant conditions it should be as close to unity as possible.

Various values of Z_f are listed in Table 3-1 from which we can see that Al/SiO$_2$ is a good match ($Z = 1.08$). Gold is frequently used and has an acceptable value of $Z = 0.381$. When additional films are deposited, such as polymer film on top of the gold electrode, the whole structure must be treated as a multiple resonator in which the reflection and/or refraction of acoustic energy occurs at each interface. Even an approximate treatment of the multiple resonator is difficult, because densities as well as thicknesses and shear moduli of the individual layers must be known. Mass loading of multiple structures often becomes too high, in which case the crystal ceases to oscillate. The value of Z affects the resilience of the whole assembly, which is most stable for $Z \sim 1$.

The response of a piezoelectric sensor to the introduction of a gas at different pressures, demonstrated as a relative change of frequency, is due

Table 3-1. Selected Values of Acoustic Impedances
of Materials Used in Mass Sensors[a]

Material	Formula	Shear-mode acoustic impedance (10^6 kg s m^{-2})
Aluminum	Al	8.22
Aluminum oxide	Al_2O_3	24.6
Chromium	Cr	29.0
Copper	Cu	20.3
Gold	Au	23.2
Graphite	C	2.71
Indium	In	10.5
Nickel	Ni	26.7
Palladium	Pd	24.7
Platinum	Pt	36.1
Quartz	SiO_2	8.27
Silicon	Si	12.4
Silver	Ag	16.7

[a] Selected values from Ref. 7 reproduced with permission of Elsevier.

to three effects: the hydrostatic effect (p), the impedance effect (x), and the sorption effect (m):

$$-\Delta f/f = (\Delta f/f)_p + (\Delta f/f)_x + (\Delta f/f)_m \qquad (3\text{-}17)$$

It is, of course, the last effect—selective adsorption or absorption of the species of interest—which is most important for chemical sensing, while the first two effects can be viewed as nonspecific interference. There are, however, situations in which those two effects cannot be ignored. For instance, equation (3-17) for an AT-cut crystal at 50 °C has the form

$$-\Delta f/f_c = 1.35 \times 10^{-9} P_{(torr)} - 7.2 \times 10^{-9}(\pi f \rho_g \eta_g)^{1/2} - 2.26 \times 10^{-6} fm$$

$$(3\text{-}18)$$

where P is gas pressure, ρ_g and η_g are gas density and viscosity, respectively, m is the mass in g cm^{-2}, and f_c is the crystal frequency. Thus for a 5-MHz AT-cut quartz crystal, the change of frequency upon transition from air at 1 atm to vacuum (about 10^{-6} mm) is -5 Hz. This change is due only to the hydrostatic effect and to the increased friction which the crystal experiences in air.

3.1.2. Piezoelectric Oscillators in Liquids

One of the first and most elegant applications of a quartz microbalance is monitoring thickness during electroplating of gold. The liquid represents an additional mass load, so coupling of the elastic shear wave to the fluid must be considered.[4] In the liquid medium shear motion of the crystal causes the adjacent layer of the solvent (water) to move along. The next, more loosely bound layer of water moves with a certain amount of "slip," which corresponds to a phase shift, etc., for the third and subsequent layers. Thus a wave in a perpendicular (longitudinal) direction is established and dissipates energy to the liquid. If only one side of the crystal is coupled to the liquid, a standing wave is formed. Acoustic attenuation in liquids is attributable to viscous loss which, for Newtonian fluids, results in the damped wave equation

$$D = D^0 \exp[(-kx) \, i(\omega t - \omega x/v)] \tag{3-19}$$

where D is the particle displacement, ω the angular frequency, v the particle velocity, x the distance in the normal direction, and k is a decay constant equal to $(\omega/2v)^{1/2}$, v being the kinematic viscosity. The damping distance (wave propagation) is on the order of micrometers (Figure 3-5) while the frequency shift is

$$\Delta f = f_0^{3/2} (v/\pi \mu_q \rho_q)^{1/2} \tag{3-20}$$

The effect of the microscopic kinematic viscosity at the crystal–liquid interface is particularly important, as seen from the frequency shift due to the increased density of the glucose solution (Figure 3-6). On the other hand, in the case where the surface energy of the interface changes due, for

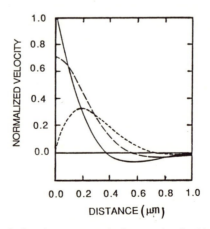

NORMALIZED VELOCITY

DISTANCE (μm)

Figure 3-5. Coupling of the shear-wave velocity to the liquid over a period of one oscillation: (——) maximum surface velocity, (----) zero surface velocity, and (— —) intermediate surface velocity. (Reprinted from Ref. 4 with permission of the American Chemical Society.)

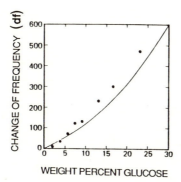

Figure 3-6. Effect of kinematic viscosity. Calculated (——) and measured (•) change of frequency with glucose concentration for a 5 MHz quartz crystal. (Reprinted from Ref. 4 with permission of the American Chemical Society.)

instance, to adsorption, energy coupling is directly affected and results in a shift of the resonant frequency. This shift is not related to the added mass. Thus the effect of "microviscosity" in condensed media has the effect on frequency comparable to or larger than the frequency shift caused by the addition of mass [see equations (3-12)–(3-14)]. This interference has the same origin as the frictional interference observed with the oscillating

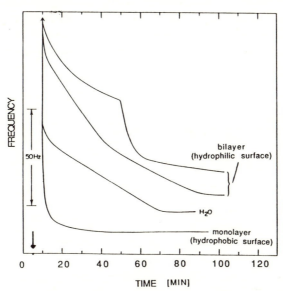

Figure 3-7. Effect of hydrophobicity of the surface on the relative frequency of a 5 MHz crystal. Stearic acid was used to form the mono- and bilayer. The arrow indicates the point of addition of water. (Reprinted from Ref. 5 with permission of the American Chemical Society.)

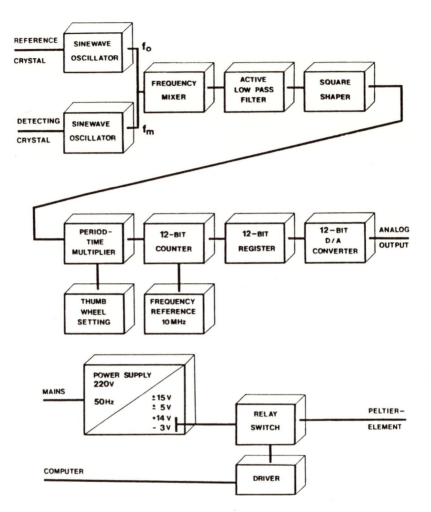

Figure 3-8. Block diagram of the dual-crystal halothane sensor. (Reprinted from Ref. 6 with permission of Elsevier.)

crystal in gas [see equation (3-18)] except that it is much more severe. It has been shown[5] that if the chemical nature of the interface changes, the frictional effects in aqueous solutions dominate the response of the oscillating crystal over any possible mass effects (Figure 3-7). This is a rather unfortunate situation, because it makes the realization of the combination of the highly sensitive microbalance with the highly selective immunochemical reactions improbable. On the other hand, when the chemical composition of the interface does not change (e.g., electroplating of gold or increasing concentration of glucose), the frictional losses are constant and the resonant frequency of the crystal follows the Sauerbrey equation. Such an application, however, cannot be regarded as true chemical sensing, because the chemical composition of the sample remains essentially constant.

Parameters which adversely affect the performance of the piezoelectric sensor include excessive mass loading due to the mass of the chemically selective layer itself, the built-in stress in the selective layer, and the stress caused by the electrodes (clamping). If the chemical species or environment causes changes in the acoustic impedance, then the resonant frequency is again affected. Many of these problems can be eliminated by use of a reference crystal (differential measurement).[6] A schematic diagram of such a microprocessor-controlled, dual-crystal instrument is shown in Figure 3-8.

3.1.3. Aerosols and Suspensions[7]

Piezoelectric microbalances can be used for monitoring heterogeneous samples such as aerosols and suspensions. Mass increase due to impacting *and* sticking particles (liquid–aerosol or solid–suspension) can be treated by the Sauerbrey equation (3-14). In that case the particle diameter is important: increased diameter increases the contact area, but increased mass of the particle increases the inertial forces trying to dislodge it from the surface. Thus there is an optimum particle diameter which is inversely proportional to the frequency of the crystal:

$$D_{crit} = \beta f_0^{-1} \qquad (3-21)$$

where the parameters included in the factor β are the particle density, adhesion, and power dissipated by the crystal. Hence the variation in frequency provides some means for discrimination based on particle size. The sensitivity decreases for particles with diameter D greater than $2\ \mu m$ and no sensitivity is obtained for particles of diameter larger than about $20\ \mu m$. In these measurements it is necessary to compensate for changes in humidity. The limit of mass loading is approximately $400\ \mu g$. Another factor which affects the response is the lateral particle distribution.

The "stickiness" of the crystal can be realized either chemically or electrostatically (Figure 3-9). In the latter case the crystal is one of the electrodes to which a voltage is applied. Grease has been used for chemical adhesion coatings of polysiloxanes or silicone.

The concentration of aerosol is usually expressed as mass of liquid

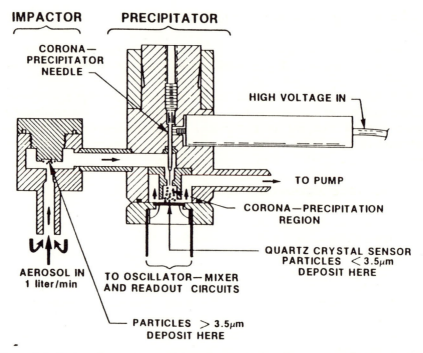

Figure 3-9. Design of a quartz crystal detector with electrostatic impactor for determination of particles in aerosols and smokes. (Reprinted from Ref. 7 with permission of Elsevier.)

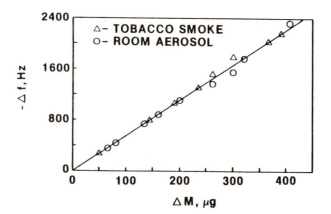

Figure 3-10. Frequency shift of a 1.6 MHz crystal of 1.06 cm² area in tobacco smoke and an aerosol. (Reprinted from Ref. 7 with permission of Elsevier.)

phase per sample volume. The collection efficiency of the sensor is defined as

$$E_c = \Delta M / M \qquad (3\text{-}22)$$

where ΔM is the mass deposited on the crystal. Assuming $E_c = 1$, a mass detection limit of 1 μg m^{-3} sample can be obtained. The microbalance has been used for assaying the particulates in tobacco smoke and in various aerosols (Figure 3-10).

3.2. Surface Acoustic Wave Sensors

In 1885 Rayleigh predicted that surface acoustic waves (SAW) could propagate along a solid surface in contact with a medium of low density, such as air (Figure 3-11).[8] These waves, which are sometimes also called Rayleigh waves, have considerable importance in areas as diverse as structural testing, telecommunications, and signal processing. An exact description of the physical processes involved in the generation, propagation, and detection of SAWs is complex and far beyond the scope of this book.[8] Nevertheless, the potential of SAW devices as chemical sensors is considerable, despite the fact that the relationships between the output signal of the SAW device with its associated electronics and the added mass are

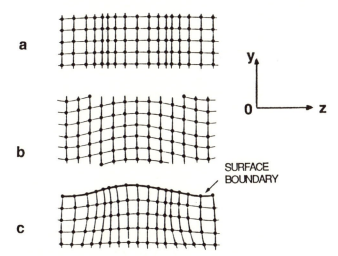

Figure 3-11. Comparison of (a) longitudinal (bulk), (b) y-polarized shear (bulk), and (c) y–0–z polarized Rayleigh (SAW) waves, all propagating in the z-direction. (Reprinted from Ref. 8 with permission of Wiley.)

mostly empirical. Subsequently we shall concentrate on the physical processes involved in the operation of these sensors.

In an isotropic medium, SAW represents a combined longitudinal and shear motion of the lattice (Figure 3-11) in the y–0–z plane, called the *saggital plane*. In anisotropic media, in certain multilayer structures and at some interfaces the SAW velocity exceeds that of the shear wave and the energy leaks continuously from the surface to the bulk of the material. In such a case we talk about pseudo- or "leaky" waves. Various energy-loss mechanisms operate on the propagating SAW: *scattering loss* due to the finite-size grains, *thermoelastic loss* due to the nonadiabatic behavior of the acoustic conductor, *viscous loss* caused by dissipation of energy in the direction normal to the liquid–solid interface, and *hysteresis absorption* due to irreversible coupling of SAW energy to the adsorbate. Addition of mass during the sensing step may change the magnitude of one or more of these mechanisms and leads to a variation in the amplitude A, or in a frequency shift Δf, a relative frequency shift $\Delta f/f$, or a phase shift $\Delta\phi$. Any of these parameters can be measured and expressed as a function of concentration.

A surface acoustic wave sensor is a transmission (delay) line in which an acoustic (mechanical) wave is piezoelectrically generated in one oscillator (transmitter, generator), propagated along the surface of the substrate, and then transformed back to an electrical signal in the receiving oscillator (Figure 3-12). From the standpoint of chemical sensing, the

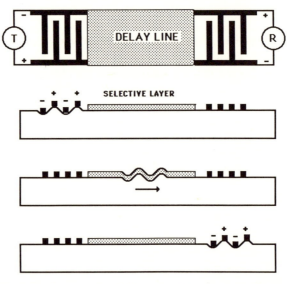

Figure 3-12. Schematic diagram of a SAW sensor with transmitter T, receiver R, and the chemically selective layer deposited on the delay line.

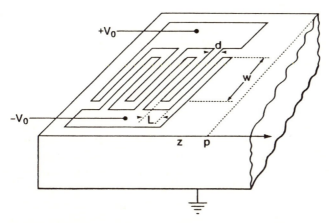

Figure 3-13. A uniform transducer (transmitter or receiver) with design parameters d, electrode width; w, aperture; and L, spatial period. For design purposes the observation line is located at distance $z = p$. (Reprinted from Ref. 8 with permission of Wiley.)

analytical information is obtained from the interaction of the sample with the traveling wave in the region of the delay line. The transmitter and generator are usually interdigitated piezoelectric transducers[9] (Figure 3-13). Usual substrate materials are ST-cut (stress- and temperature-compensated) quartz, or $LiNbO_3$ which has a high piezoelectric coefficient. The area between the transmitter and receiver, the delay line, does not have to be made from the same material; as a matter of fact it need not even be piezoelectric. In that case acoustic couplings are used to transfer the energy from the piezoelectric transmitter/receiver to and from the delay line. A combination of all these factors results in a perplexing number of possibilities for the design of SAW sensors and for the modes in which the output signal can be obtained. The integrated circuit techniques can be used to fabricate these devices. It is therefore convenient to represent the SAW sensor by an equivalent electrical circuit and perform some optimization beforehand (Figure 3-15). The optimization parameter is the energy loss, which should be minimized. Another important factor to be taken into consideration is the size of the sensor. The scaling relationships which apply to the SAW sensors are summarized in Table 3-1. It is clear that a compromise frequency must be found in order to achieve the optimum operational characteristics.

Most SAW sensors employ a dual arrangement: a sensor delay line and the reference line combined on one substrate with one generator and two receivers[10] (Figure 3-14). The phase shifts in the reference (ϕ_r) and sensing (ϕ_s) lines are given by

$$\phi_r = 2\pi f l_r / v_r + \phi_E \qquad (3-23)$$

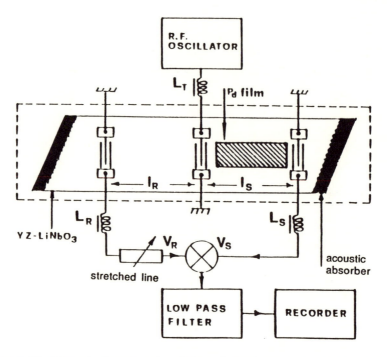

Figure 3-14. Schematic diagram and equivalent electrical circuit of a dual-line hydrogen SAW sensor. (Reprinted from Ref. 10 with permission of Elsevier.)

and

$$\phi_s = 2\pi f l_s / v_s \tag{3-24}$$

where ϕ_E is the phase shift introduced by the compensating (stretched) line, l_r and l_s are the lengths of the reference and sensing channels, respectively, while v_r and v_s are the acoustic velocities in the two channels. The output of the sensor, V_0, is then given by

$$V_0 = V_M \sin(\phi_r - \phi_s) \tag{3-25}$$

where V_M is the maximum output occurring at $\phi_r - \phi_s = k\pi/2$ (for $k = 1, 3, ..., 2n - 1$). The stretched line is used to set the phase-shift difference for zero concentration. Equations (3-22)–(3-25) yield

$$V_0 = V_M \sin[2\pi f(l_r/v_r - l_s/v_s) + \phi_E] \tag{3-26}$$

The addition of the mass due to interaction with the gas causes the change of acoustic velocity v_s in the sensing line. Therefore, the sensitivity

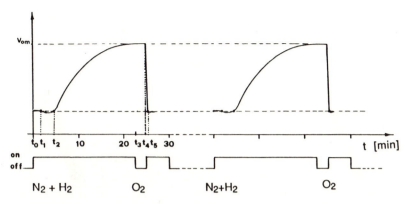

Figure 3-15. Change in output voltage V_0 due to injections of 1% H_2 in N_2 in the presence of oxygen. The flow rate is 13.8 ml s^{-1} and the O_2 flow rate is 2.5 ml s^{-1}. (Reprinted from Ref. 10 with permission of Elsevier.)

of the device is obtained by differentiating equation (3-26) with respect to C_{gas}. This procedure leads to

$$S = \partial V_0/\partial C_{gas}$$
$$= -2\pi V_M f l_s/v_s^2 (\partial v_s/\partial C_{gas}) \cos[2\pi f(l_r/v_r - l_s/v_s) + \phi_E] \qquad (3\text{-}27)$$

By using the stretched line ϕ_E, the phase shift can be adjusted in such a way that the argument in the cosine is zero, in which case equation (3-27) reduces to

$$S = \partial V_0/\partial C_{gas} = -2\pi V_M f l_s/v_s^2 (\partial v_s/\partial C_{gas}) \qquad (3\text{-}28)$$

The temperature and pressure also affect the SAW velocities. The pressure (S_p) and temperature (S_T) sensitivities are respectively

$$S_p = \partial V_0/\partial p = 2\pi V_M f \left(\frac{l_s}{v_s^2} \frac{\partial v_s}{\partial p} - \frac{l_r}{v_r^2} \frac{\partial v_r}{\partial p} \right) \qquad (3\text{-}29)$$

and

$$S_T = \partial V_0/\partial T = 2\pi V_M f \left(\frac{l_s}{v_s^2} \frac{\partial v_s}{\partial T} - \frac{l_r}{v_r^2} \frac{\partial v_r}{\partial T} \right) \qquad (3\text{-}30)$$

Hence a prudent choice of l_s and l_r can minimize the pressure and temperature dependence of these devices.

An example of the above device is the hydrogen SAW sensor shown in Figure 3-14, where the area of L_s is coated with 3000 Å of Pd. The substrate is LiNbO$_3$. There are five finger pairs and the device is operated at $f_0 = 75$ MHz. The width of the acoustic beam is 100 acoustic wavelengths. The sensor responds to a step change of hydrogen concentration with rise time approximately 18 min in nitrogen and fall time 8 s in oxygen (Figure 3-15). The dynamic range is 50–10,000 ppm (Figure 3-16).

Thin layers of organic films can be used to provide chemical selectivity. As with piezoelectric crystals the acoustic properties of these films affect the performance of the SAW sensor in different ways: by the change in mass due to the adsorbed/absorbed gas or by the change in the elastic properties of the film. An electrostatic mode of coupling between the chemically selective coating and the piezoelectric delay line is also possible.[11] It may happen when the film is a semiconductor with a low density of charge carriers. As the piezoelectric wave propagates through the delay line, it induces a traveling electric wave in the adjacent film whose velocity depends on the conductivity of the film. That, in turn, can be modulated by the interaction of the film with an electron-donating or electron-accepting gas.

A comparative study of the readout options for the SAW sensor with additional film has shown that for a single SAW sensor the best signal-to-noise ratio is obtained from the amplitude measurement.[12] Voltage output related to phase shift as discussed above works well for dual delay lines.

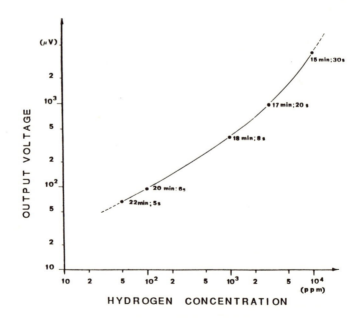

Figure 3-16. Calibration curve of the hydrogen SAW sensor shown in Figure 3-14. Parameters given by the experimental points correspond to the rise and fall time for flow rates given in Figure 3-15. (Reprinted from Ref. 10 with permission of Elsevier.)

There are inherent advantages in measuring the change in resonant frequency. The frequency shift due to a deposited film of *low elastic modulus* μ is

$$\Delta f = (k_1 + k_2) f_0^2 t_f \rho - k_2 f_0^2 t_f [4\mu(\chi + \mu)/V_r^2(\chi + 2\mu)] \quad (3\text{-}31)$$

where k_1 and k_2 are material constants of the SAW substrate, t_f is the film thickness and ρ its density, while χ is the Lamé constant and μ the elastic modulus of the film. Thus, the film mass per unit area is $t_f \rho$. If the chemical interaction does not alter the mechanical properties of the film, the second term in equation (3-31) can be neglected and the frequency shift Δf is due exclusively to the added mass,[12]

$$\Delta f = (k_1 + k_2) f_0^2 t_f \rho \quad (3\text{-}32)$$

Thus addition of a 1-μm-thick polymer film coated onto a quartz SAW sensor operating at 31 MHz will cause a $\Delta f = -130$ kHz shift.

3.2.1. Plate Mode Oscillators

In the SAW devices discussed above, the acoustic wave propagates in a slab of material whose thickness is infinitely larger than the wavelength of the propagating wave. When the thickness of the plate is reduced so that it becomes comparable to λ, the whole plate becomes involved in the periodic motion and a symmetric and antisymmetric Lamb wave is created (Figure 3-17). This happens in plate-mode oscillators (PMO) with thicknesses of a few microns.

There are certain performance aspects which make the PMO potentially attractive as chemical sensors. Both surfaces contribute to the signal,

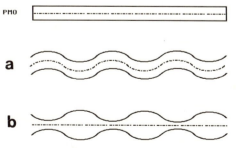

Figure 3-17. Lamb waves propagating in a thin plate-mode oscillator: (a) symmetric and (b) antisymmetric waves.

so the sensitivity is higher than for the corresponding SAW device. The most important advantage follows from the fact that the velocity of the lowest order of the antisymmetric mode is much slower than that of the corresponding SAW oscillator. This may be important for potential applications to mass sensing in liquids, which is problematic with high-frequency SAW devices. As the frequency of the Lamb wave decreases below the velocity of the compressional wave in the liquid, the energy loss in the normal direction also decreases.

There are also some practical problems. The fragility of the thin plate and the sensitivity to external pressure being most serious. The Lamb wave has also a higher-frequency dispersion than the corresponding surface acoustic wave. Analysis of the performance as well as the details of fabrication can be found in the original paper.[13]

3.3. Selective Coatings

Piezoelectric crystals have been used as thickness monitors. They are not really true chemical sensors, because the density (and composition) of the deposited film must be known. The information sought is the *thickness* of the layer [equations (3-12) and (3-32)]. The sensors are used in the monitoring of electroplating, chemical vapor deposition, sputtering, radio-frequency plasma deposition, etc. The usual range of thicknesses is from submonolayer to 2000 Å. A great advantage of these sensors is that they can be used for monitoring in vacuum.[14]

For a true chemical sensing application a chemically selective layer must be deposited on the mass sensor. The requirements pertaining to the mechanical and chemical properties of such layers are similar for the piezoelectric microbalance and for the SAW devices. One of the most attractive features of mass sensors is the ease with which the selective coatings can be applied to the transducer. As long as the application of the layer does not damp the oscillation of the crystal or does not quench the propagation of the SAW due to mass loading or acoustic impedance mismatch, any organic or inorganic layer can be used. This fact alone may account for the large number of papers describing mass sensors for various species.[3,15] The layers are applied by dip-coating, spraying, painting, sublimation, or chemical vapor deposition. On SAW, spin-coating followed by lift-off or etch patterning is commonly used. The materials used in the construction are often oxides (such as SiO_2), so it is possible to use covalent immobilization of the binding sites and thin layers directly at the crystal or transmission line surface.

The choice of coating in terms of chemical selectivity is usually gover-ned by some knowledge of the interaction of the species of interest with the

Table 3-2. Comparison of SAW and GLC Equilibrium Constants
for Sorption on Fluoropolyol[a]

Substance	log K_{SAW}	log K_{GLC}
Dimethyl methylphosphonate	6.52	7.53
N,N'-Dimethylacetamide	6.33	7.29
1-Butanol	3.83	3.66
2-Butadone	3.38	3.48
Water	3.20	2.89
Toluene	2.88	2.64
Diethyl sulfide	2.74	2.54
1,2-Dichloroethane	2.46	1.94
Isooctane	2.12	1.22

[a] Reprinted from Ref. 16 with permission of the American Chemical Society.

selective layer. Thus, information obtained by gas chromatography about the interaction of various gaseous constituents with the column materials has served as a useful guide[16] (Table 3-2). For example, a mixture of high-molecular-weight hydrocarbons (and oxyhydrocarbons) has continuously variable *polarity*. A typical example are Carbowaxes (polyethylene glycols and their esters, and related polyethers). The affinity of such materials to, for instance, hydrocarbons is only modest. It is based primarily on weak interactions (cf. Section 1.1) without any use of shape recognition. Nevertheless, in *chromatographic applications* they provide very high resolution because the equilibration process is repeated many thousands of times on the chromatographic column. This is a very important difference between *direct sensing* and *separation*. There is only one equilibration step in the sensor application, which means that the requirements on the magnitude and specificity must be higher than those for the same material used in chromatography. Moderately selective mass sensors for aromatic, aliphatic, and halogenated hydrocarbons based on these coatings have been described.[3]

As long as the effect of the chemical species on the sensor is related only to the selective addition of the mass to the selective layer, the response will be governed by some adsorption isotherm [see equations (A-27) and (A-28), and Table 1-1] in the case of surface adsorption, or by the partitioning equilibrium [see equation (1-12)] in the case of bulk interaction. In both cases the Sauerbrey equation [see equations (3-12)–(3-14)] or the SAW equation (3-32) applies. However, in using these coatings the possible change in the mechanical properties of the film due to interaction with the chemical environment (e.g., elasticity) must also be considered. In most cases these relationships are not known *a priori* and only empirical dependencies of the sensor output on the composition of the sample (calibration) can be used. Nevertheless, valuable chemical information can be obtained.[12]

Probably the most important type of piezoelectric mass sensor which operates on the principle of bulk changes in the selective layer is a humidity monitor. The best coating for this application seems to be a relatively thick (150 μm) coating of gelatine.

Another advantage of both types of mass sensor is their relatively low cost. In most cases they can be employed in some differential arrangement, in which case unwanted interferences can be reduced or eliminated by the reference device. In the case of SAW sensors the reference channel is sometimes designed directly in the device.[10] Below are examples of some selected coatings and their applications.[15]

Hydrogen Sulfide. Selective layers containing WO_3, lead, silver, or copper acetate have been used. The cations of these metals have high affinity for H_2S. The obvious interferences include mercaptans, which also have high affinity for heavy metals.

Carbon Dioxide (dissolved). In a way this sensor is a "mass analog" of the Severinghaus electrode. The crystal is coated with didodecylamine —a base is separated from the sample by a hydrophobic membrane (such as Teflon, PVC, or SR). When the acidic gas penetrates through the membrane it reacts with the basic coating, thus increasing the mass of the crystal. Expected interferences include all acidic gases, such as acetic acid, H_2S, SO_2, HCN, and HCl. The reference crystal is sealed and provides compensation for variations in temperature.

Mercury. Gold electrodes on a crystal are used as the selective layer. Gold has high affinity for metallic mercury. Expected interferences are halogens, H_2S, and HCN.

Halothane.[6] In this application two matched quartz crystals are used: one coated with silicone oil and the other bare. The frequencies of the two crystals are mixed and the beat frequency (f_d) selected by the low pass filter. The beat frequency difference is inversely proportional to the beat period T,

$$\Delta f_d = 1/T \tag{3-33}$$

A change due to adsorption is $\Delta f_{gas} = -\Delta f_d = f_d^2 \, \Delta T$. The dynamic range for halothane is 0.5–100 ppm.

Glossary for Chapter 3

C_f	Crystal constant
C_0	Stray capacitance
D	Particle displacement
f	Resonant frequency
f_d	Beat frequency
L	Electrical inductance
l	Length
M	Mass
m	Mass per unit area
P	Gas pressure
R	Electrical resistance
T	Beat period
t	Thickness
v	Shear velocity
Z	Acoustic impedance
χ	Lamb constant
ϕ	Phase shift
λ	Wavelength
μ	Shear modulus
ρ	Density
ω	Angular frequency
ν	Kinematic viscosity

References for Chapter 3

1. A. W. Warner, Microweighing with the quartz crystal oscillator—theory and design, in: *Ultra Micro-Weight Determination in Controlled Environments* (S. P. Wolsky and E. J. Zdanuk, eds.), Interscience, New York, 1969.
2. W. H. King, Jr., *Anal. Chem.* 36 (1964) 1735.
3. G. G. Guilbault, Applications of quartz crystal microbalances in analytical chemistry, in: *Methods and Phenomena*, Vol. 7 (C. Lu and A. W. Czaderna, eds.), Elsevier, Amsterdam, 1984.
4. K. K. Kanazawa and J. G. Gordon, *Anal. Chem.* 57 (1985) 1770.
5. M. Thompson, C. L. Arthur, and G. K. Dhaliwal, *Anal. Chem.* 58 (1986) 1206.
6. A. Kindlund, H. Sundgren, and I. Lundstrom, *Sensors and Actuators* 6 (1984) 1.
7. M. H. Ho, Application of quartz crystal microbalances in aerosol mass measurement, in: *Methods and Phenomena*, Vol. 7 (C. Lu and A. W. Czaderna, eds.), Elsevier, Amsterdam, 1984.
8. V. M. Ristic, *Principles of Acoustic Devices*, Wiley, New York, 1983.
9. W. J. Ghijsen and A. Venema, *Sensors and Actuators* 3 (1982/83) 51.
10. A. D'Amico, A. Palma, and E. Verona, *Sensors and Actuators* 3 (1982/83) 31.
11. A. J. Ricco, S. J. Martin, and T. E. Zipperian, *Sensors and Actuators* 8 (1985) 319.
12. H. Wohltjen and R. Dessy, *Anal. Chem.* 51 (1979) 1465.

13. R. M. White, P. J. Wicher, S. W. Wenzel, and E. T. Zellers, *IEEE Trans. Ultrason. Dev. Ferroelectr. Freq. Control* UFFC-34 (1987) 162.

14. H. K. Pulker and J. P. Decosterd, Applications of quartz crystal microbalances for thin film deposition process control, in: *Methods and Phenomena*, Vol. 7 (C. Lu and A. W. Czaderna, eds.), Elsevier, Amsterdam, 1984.

15. J. F. Alder and J. McCallum, *Analyst* 108 (1983) 1169.

16. J. W. Grate, A. Snow, D. S. Ballantine, Jr., H. Wohltjen, M. H. Abraham, R. A. McGill, and P. Sasson, *Anal. Chem.* 60 (1988) 869.

4

Electrochemical Sensors

Electrochemical sensors are the largest and oldest group of chemical sensors. Many members of this group have reached commercial maturity while many are still in various stages of development. They will be discussed within the broadest framework of electrochemistry, the interaction of electricity and chemistry. Sensors as diverse as enzyme electrodes, high-temperature oxide sensors, fuel cells, surface conductivity sensors, etc., will be included. They are divided by their mode of measurement into *potentiometric* (measurement of voltage), *amperometric* (measurement of current), and *conductimetric* (measurement of conductivity) sensors.

Electrochemistry implies the transfer of charge from an electrode to another phase, which can be a solid or a liquid sample. During this process chemical changes take place at the electrodes and the charge is conducted through the bulk of the sample phase. Both the electrode reactions and/or the charge transport can be modulated chemically and serve as the basis of the sensing process. There are some common rules which apply to all electrochemical sensors, the cardinal one being the requirement of a *closed electrical circuit*. This means that at least two electrodes constitute an *electrochemical cell*. From a purely electrical point of view we have a sensor electrode and a *signal return*.

This requirement does not necessarily mean that a DC electric current will flow in a closed circuit. Obviously, if we consider an ideal capacitor C in series with a resistor R (Appendix D), a DC voltage V will appear across the capacitor but there will be zero DC current [see equations (D-1) and (D-2)]. On the other hand, if an AC voltage is applied then a displacement current will flow. An "open" electrical circuit (such as an "electrochemical cell" consisting of only one electrode!) simply means that a small and undefined capacitor (the missing electrode) has been placed in series in such a circuit. Its performance can be predicted from equation (D-3): it will readily respond to high-frequency electrical fluctuations (e.g., noise) but no information which depends on the DC behavior can be obtained. This is not necessarily a problem. For example, an antenna transmitting at GHz frequency can be regarded as one electrode while the receiver, the "other electrode," does not even have to be defined (in the extreme it can be Earth). If a dielectric object is placed in the proximity of the transmitter, the amount of reflected radio-frequency energy depends on the average dielectric constant of such an object. This is the basis of one of the most accurate, continuous humidity sensors, because the dielectric constant of the sample is a strong function of the water content. Admittedly, this electrochemical sensor hardly lies within the scope of this book, but it clearly illustrates the point that the *frequency* of the applied signal is an important parameter in considering the continuity of the electrical circuit.

Another important general aspect of electrochemical sensors is that the charge transport within the transducer part of the sensor and/or inside the supporting instrumentation, which is part of the whole circuit, is always electronic. On the other hand, the charge transport in the sample can be electronic, ionic, or mixed. In the latter two cases *electron transfer* and sometimes (for net nonzero DC current) *electrolysis* take place at the electrode–sample interface and its mechanism becomes one of the most critical aspects of the sensor performance. In some traditional electro-chemical texts[1-4] the processes taking place at the interface are classified under the heading *electrodics*, while the bulk charge transport properties can be found under *ionics*.

What happens inside an electrochemical cell? The number of varia-tions in the electrochemical cell resulting from the combination of phases and their modes of conduction, frequencies, and temperatures is large consequently, it is not possible to construct a current–voltage characteristic which would be both sufficiently representative and general at the same time. We shall therefore examine the case of a $0.1 \, M \, H_2SO_4$ aqueous solution (ionically conducting phase) and a pair of electrodes of unequal size (Figure 4-1). The smaller will be called the *working* (or *indicator*) electrode (W) and the larger, the *auxiliary* electrode (Aux). They are con-nected to the signal source S (current or voltage) and a meter M (voltage or current). It is assumed that the voltage is an independent variable which is applied to this cell such that the working electrode is negative with respect to the auxiliary electrode. At first no current flows through the cell until a decomposition potential E_{d-} is reached, after which the current begins to flow. The polarity of the applied voltage, and the scale of the voltage axis in Figure 4-1, are always referred to the smaller of the two electrodes, as will be explained in detail later. Two chemical processes take place at the electrodes. At the working electrode (in this case the cathode) electrons will be transferred from the electrode to the adjacent hydrogen

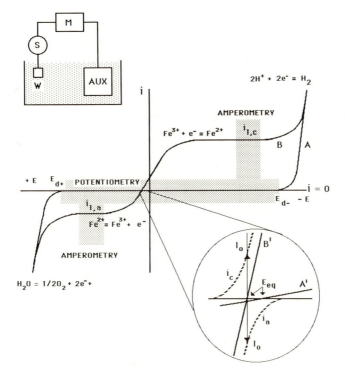

Figure 4-1. General electrochemical experiment.

ions H^+ at the surface of the electrode (reduction) according to the chemical reaction

$$2H^+ + 2e^- = H_2 \tag{4-1}$$

At the anode (the larger, auxiliary, electrode) the opposite electrochemical reaction will take place at potentials more positive than the decomposition potential E_{d+},

$$H_2O = \tfrac{1}{2}O_2 + 2e^- + 2H^+ \tag{4-2}$$

Thus, for a transfer of two electrons in the external circuit one molecule of water is decomposed (electrolyzed) to half a mole of oxygen and one mole of hydrogen. We must realize, however, that the positive half of curve A can be recorded on the same current scale *only* if we change the polarity of the applied voltage to the working electrode (from 0 to $+2$ V). Then, of course, the electrons are withdrawn from the solution at that electrode (oxidation) according to equation (4-2). The voltage between the

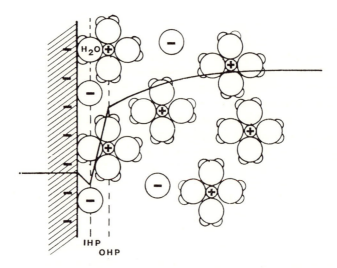

Figure 4-2. Model of a double-layer capacitor and the potential profile. The anions are specifically adsorbed despite the negative charge on the electrode. Specifically adsorbed ions are equivalent to surface states at semiconductor surfaces. (Reprinted from Ref. 24 with permission of the American Chemical Society.)

two decomposition potentials represents a voltage window in which the electrochemical measurements can be made. From the electrical point of view the working electrode is a capacitor with one plate being the electrode and the other the ionic charge accumulated at the solution side of the interface. It is called a double-layer capacitor (Figure 4-2). The energy of this capacitor depends on the voltage applied to it. The dielectric breakdown of the "dielectric" in this capacitor* occurs either at the positive (E_{d+}) or negative (E_{d-}) decomposition potential.

A mixture of $Fe^{II}(NH_4)_2(SO_4)_2$ and $Fe^{III}NH_4(SO_4)_2$ is now added in approximately 10^{-3} M concentration and the current–voltage curve again recorded (curve B). The reason why it looks different from curve A is that the added redox couple depolarizes (shorts out) the double-layer capacitor and electrons can jump across the interface at much lower energy (voltage) than before. Thus the energy of the capacitor has been used to oxidize or reduce the added *depolarizer* in the solution. The equivalent electrical circuit corresponding to this situation is shown in Appendix D (Figure D-1).

We now examine qualitatively the shape of the current–voltage curve B. First, we note that it intercepts the potential axis at the equilibrium

---

* The notion of the capacitor is helpful in modeling this situation with equivalent electrical circuits. However, it must be remembered that the value of this capacitor is a complex function of potential.

potential E_{eq}. As the potential of the working electrode is made more negative with respect to E_{eq}, the current first increases due to the reduction of Fe^{3+} ions and then levels off at the limiting-current plateau. When it reaches E_{d-} it begins to increase again as the reduction of protons starts to contribute to the total cathodic current. Similarly, when we start at the equilibrium (zero-current) potential and move in the positive direction, the ferrous (Fe^{2+}) ions are oxidized and this reaction merges with the oxidation of water beyond the positive decomposition potential. It will be seen later that the two limiting currents, $i_{l,c}$ and $i_{l,a}$, in the plateau regions are used as the signal in amperometric sensors. Two of the three principal measuring domains of electrochemical sensors are defined in Figure 4-1 as shaded areas: potentiometric at the zero-current line and amperometric at either the cathodic or anodic limiting-current plateau. The conductimetric measurement can be carried out at zero DC current, in which case a small-amplitude periodic signal is applied such that the integrated net charge is zero. The measurement can also be conducted at a finite DC current. In that case, however, the electrochemical reactions takes place at the electrode interface with the resulting net chemical changes governed by Faraday's law

$$\text{mole} = Q/nF \tag{4-3}$$

where Q is the total charge passed through the electrode during the experiment, n the number of electrons per mole, and F is the Faraday constant. Such changes can have a cumulative effect on the measurement, particularly when the products of the electrochemical reaction are not removed sufficiently rapidly from, and the electroactive species supplied to, the interface. This condition is called *electrochemical polarization* and is a consequence of Faraday's law.

The overall current–voltage relationship is complex and differs for various conditions. It is affected mainly by the nature and concentration of the electroactive species, by the electrode material, and by the mode of mass transport. The general relationship which describes the current as a function of applied voltage is

$$i = nFAk_0\{C_O(0, t)\exp[-\alpha nF(E - E_0)/RT]$$
$$- C_R(0, t)\exp[(1 - \alpha)nF(E - E_0)/RT]\} \tag{4-4}$$

where $C_O(0, t)$ and $C_R(0, t)$ are the surface concentrations of the oxidized and reduced form of the depolarizer, respectively, α is the symmetry coefficient, A the area of the electrode, and k_0 is called the heterogeneous rate constant. The symmetry coefficient does not have any direct consequences for electrochemical sensors. A discussion of its meaning and importance can be found in electrochemical textbooks.[1-4] Its value lies typically between 0.3 and 0.7. The total current in equation (4-4) can be decomposed to its cathodic (i_c) and anodic (i_a) components,

$$i_c = nFAk_0 C_O(0, t) \exp[-\alpha nF(E - E_0)/RT] \qquad (4-5)$$

and

$$i_a = nFAk_0 C_R(0, t) \exp[(1 - \alpha) nF(E - E_0)/RT] \qquad (4-6)$$

where

$$i = i_c - i_a \qquad (4-7)$$

The general properties of equation (4-4) can now be examined qualitatively. For an increasingly more negative potential applied to the electrode, the cathodic contribution to the overall current increases exponentially [see equation (4-5)]. It levels off as it reaches the limiting plateau region in which the mass transport of the oxidized species to the electrode surface becomes the current limiting factor. Simultaneously, the exponential term governing the anodic contribution [equation (4-6)] becomes smaller. A symmetrical argument exists for the increasingly positive potential at the working electrode, in which case the dominating current in equation (4-4) is the anodic current. The fact that in the neighborhood of the equilibrium potential *both cathodic and anodic currents* contribute to the overall current in comparable amounts is shown in Figure 4-1 (insert). These partial contributions are shown as dashed lines (for the Fe^{3+}/Fe^{2+} redox couple) while the net current is shown as the solid line (B′).

At equilibrium ($E = E_{eq}$) the net current is zero by definition, and because there is no electrolysis taking place at the electrode the surface concentrations are equal to the bulk concentrations C_O^* and C_R^*, respectively. Equations (4-5)–(4-7) yield

$$C_O^*/C_R^* = \exp[nF(E_{eq} - E_0)/RT] \qquad (4-8)$$

or

$$E_{eq} = E_0 + \frac{RT}{nF} \ln \frac{C_O^*}{C_R^*} \qquad (4-9)$$

which is the Nernst equation defined from kinetic considerations.* Later (in Section 4.1) we shall derive the same relationship from the ther-

---

* In reaction kinetics it is customary to use concentrations instead of activities in the rate equations (cf. Section 1.1).

modynamic point of view. It is noteworthy that for equal concentrations of the oxidized and reduced species the logarithmic term in equation (4-9) vanishes, and the equilibrium potential becomes the standard potential E_0.

A glance at the insert in Figure 4-1 reveals that under these conditions the two currents are equal in magnitude but opposite in sign. This means that at zero current there is an *exchange current* i_0 passing through the interface and is shown as two vertical arrows (insert),

$$i_c = i_a = i_0 \qquad (4\text{-}10)$$

Clearly, the condition of zero net current can be satisfied for any pair of equal cathodic and anodic currents. For the sake of clarity the exchange currents for the oxidation/reduction of water (curve A′) have been omitted in Figure 4-1. It is intuitively evident that the magnitude of the exchange current for the latter process is much smaller than for the ferro/ferri couple. Thus, the magnitude of the exchange current is responsible for the shape of the current–voltage curve, as can be seen from Figure 4-1 by comparing curves A and B.

We can also select the concentrations of the oxidized and reduced species to be equal, $C_O^* = C_R^* = C$ [e.g., $E = E_0$, according to equation (4-9)], in which case equations (4-5) and (4-6) yield the expression for the exchange current i_0 in terms of the *standard heterogeneous rate constant*,

$$i_c = i_a = i_0 = nFAk_0C \qquad (4\text{-}11)$$

This equation establishes the relationship between the rate constant k_0 and the exchange current, which can be obtained experimentally. Equation (4-11) is valid only under standard conditions. For values of E close to the standard potential E_0, the exponential terms in equations (4-5) and (4-6) can be expanded employing the approximation $e^x \approx 1 + x$. After substitution in equations (4-7) and (4-11) we obtain

$$i = i_0 \frac{nF}{RT}(E - E_0) \qquad (4\text{-}12)$$

Differentiation of this equation with respect to potential yields

$$di/dE = i_0 nF/RT \qquad (4\text{-}13)$$

which is the slope of the current–voltage curve at equilibrium. Its inverse has dimensions of resistance and is the *charge-transfer* resistance R_{ct},

$$dE/di = R_{ct} = RT/nFi_0 \qquad (4\text{-}14)$$

Both the charge-transfer resistance and the exchange-current density are critically important parameters for the operation of most electrochemical sensors. The effect of the value of $i_0(k_0)$ on the shape of the current–voltage curve is shown in Figure 4-3. For an applied potential sufficiently far from the standard potential, the cathodic [equation (4-5)] or anodic [equation (4-6)] current always reaches its mass-transport limited value. This fact is particularly important for the operation of amperometric sensors.

From the electrical point of view, the electrode with a fast equilibrium process (such as one with a high value of the heterogeneous rate constant and a high value of the exchange current) represents a good ohmic contact. Now we must return to our experiment and ask: Why did we choose a small working electrode and a large auxiliary electrode?

Without further explanation it was stated that the size of the electrode determines the polarity, as shown on the potential axis in Figure 4-1. In order to understand this argument it helps to replace the two electrodes in Figure 4-1 by equivalent charge-transfer resistances R_{ct}. The exchange current [equation (4-11)] and therefore the charge-transfer resistance [equation (4-14)] are directly proportional to the area A, so the smaller electrode will have a higher resistance, all other parameters remaining

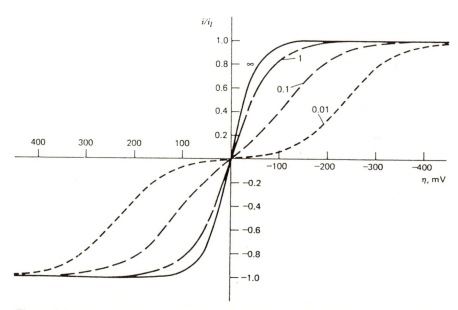

Figure 4-3. Effect of the exchange-current density on the shape of the current–voltage curves. Calculated curves are for $\alpha = 0.5$, $T = 298$ K, and $n = 1$. (Reprinted from Ref. 4, p. 110 with permission of Wiley.)

unchanged. Therefore, the electrochemical processes taking place at the smaller electrode will determine the absolute value of the current flowing through the circuit. The auxiliary electrode will begin to interfere only if its charge-transfer resistance, which is a function of the area and of k_0, becomes comparable to that of the working electrode. From this point of view it is correct to speak of *current density* and *exchange current density* rather than of current and exchange current when discussing the current–voltage behavior of any system. This is extremely important from the sensing standpoint, because in such a case information about the concentration of the depolarizer comes *only* from the smaller working electrode. In other words, it really does not matter where and how far away the auxiliary electrode is located as long as the working electrode is the current limiting element. Such a condition can be easily satisfied in amperometric applications of microelectrodes (cf. Section 4.2). On the other hand, in zero-current potentiometry the relative size of the two electrodes is immaterial (because the current is zero). Thus, in order to obtain useful information, the potential of the working electrode must be measured against a well-defined and stable potential of a reference electrode (cf. Section 4.1). Any inadvertently present potential within the measuring circuit can contaminate the information. For this reason, it is mandatory to place the reference electrode as close to the indicator (working) electrode as is practically possible. Thus, the source of information can be localized in amperometric (and conductimetric) measurement by choosing a small working electrode, while it cannot be localized in zero-current potentiometric measurements.

So far we have considered the electrode–solution interface from the kinetic point of view. The double-layer capacitor C_d has only been mentioned in passing and the fact that within the electrochemical window, defined by the two decomposition potentials, it behaves as a real capacitor. Both the insert in Figure 4-1 and equations (4-5) and (4-6), however, should leave no doubt that this capacitor is "leaky." This "leak" of electric charge (ions or electrons) is clearly related to the value of the charge-transfer resistance. Let us forget the real world for a moment and assume that there exists an interface across which no charge can pass. A skeptical reader could argue that one can always apply a sufficiently high voltage to break down any capacitor, but let us be optimistic for a while. Let there be no charge transfer across this hypothetical interface! Such an electrode will be called an *ideally polarized* electrode (sometimes also a "blocked electrode"). It will be a true ideal capacitor from the viewpoint of the equivalent circuit model. Its solid-state physics equivalent is an ideal Schottky barrier at the metal–semiconductor interface. The important general feature of any interface is that it distributes charge (ions) at the solution side of the interface. This distribution is governed by the existing electric field in that region. If we can control the electric field externally, such as on a capacitive interface, we can *polarize* such an electrode. This interface is reversible and can be at equilibrium; however, the equilibrium conditions include an additional parameter, charge (an additional degree of freedom), as compared to the Nernst equation (4-9), which describes the *nonpolarized* interface. It has been seen that conversion of a polarized interface to a nonpolarized interface can be effected simply by adding a suitable depolarizer to the solution, in our case the ferro/ferri couple. Thus, the nonpolarized interface at constant pressure and temperature can be described uniquely by the potential and concentration (or activity, as we shall see later) while equilibrium at the polarized interface is described by potential, concentration, *and charge*,

$$\gamma = q_i E + \sum_i \Gamma_i \tilde{\mu}_i \qquad (4\text{-}15)$$

This is the Gibbs–Lippmann equation in which q_i is the interfacial charge, E the potential, γ the interfacial energy, Γ_i the relative surface concentration, and $\tilde{\mu}_i$ is the electrochemical potential of the species adsorbed at the interface. This would be the equation to use if a sensor could be devised which would rely on measurement of chemically modulated excess charge at the interface (such as a potentiometric immunoelectrode[5]). Unfortunately, as soon as the interfacial capacitor begins to "leak" charge the equilibrium becomes defined by the Nernst equation.

This argument can best be understood by considering the equivalent electrical circuit for a *real*, as opposed to an ideal, interface (Appendix D, Figure D-1). It is composed of a parallel combination of a double-layer capacitor C_d and charge-transfer resistor R_{ct}. In such a representation the ideally polarized interface is the one with an infinitely high value of charge-transfer resistance. Any finite value of R_{ct} will cause the charge to leak across the capacitor and direct measurement of the excess charge is no longer possible. The difference between the ideally polarized interface and a nonpolarized interface is shown in Figure 4-4. Returning for a moment to the electrochemical experiment shown in Figure 4-1, we see that curve A *approaches* the behavior of an ideally polarized interface while the solution containing the fast redox couple makes the electrode behave as a nonpolarized (Nernstian) interface. It is important to realize that the polarizability and nonpolarizability of the electrode depend on its material *and* on all the species present in the solution. There is continuous transition between these two limiting cases, the transition depending on the value of the heterogeneous rate constant (exchange-current density and charge-transfer resistance).

A certain amount of selectivity can be obtained for amperometric

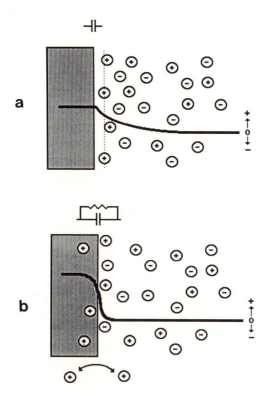

Figure 4-4. Model of an interface at equilibrium and of the potential profile. (a) Polarized (capacitive) interface governed by equation (4-15) and (b) nonpolarized (resistive) interface governed by equation (4-18). The dotted line in (a) represents the closest approach of the solvated ion. The solid phase is shown as conductive in both cases.

sensors by judicious selection of the potential of the working electrode. However, we must remember that whether or not the electron transfer takes place depends on the difference between the potential of the electrode and the equilibrium potential of the electrochemical species in the sample. At a given applied electrode potential *all* species whose potential is equal to or lower than this latter value will undergo electrolysis. Therefore, the main means of achieving chemical selectivity remains the chemically selective layer.

4.1. Potentiometric Sensors

Potentiometric measurements are conducted under conditions of zero current. Therefore the domain of this group of sensors lies at the zero-current axis (Figure 4-1). There are two types of electrochemical interface from the viewpoint of the charge transfer: ideally polarized (purely capacitive) and nonpolarized. As the name implies the ideally polarized interface is only hypothetical. Some metals (such as Hg, Au, or Pt) in contact with solutions containing only inert electrolyte (e.g., H_2SO_4) approach the behavior of such a interface. Nevertheless, even in those cases a finite charge-transfer resistance exists at such an interface and excess charge leaks across with the time constant given by the product of the double-layer capacitance and the charge-transfer resistance ($\tau = R_{ct}C_{dl}$).

With the introduction of ion-sensitive field-effect transistors (ISFET) the possibility of using the gate insulator of these devices as the active electrochemical surface became real. Experimental evidence suggests that even these insulators, when immersed in water, behave as a nonpolarized interface. Although possible in principle there are at present no chemical sensors based on a polarized interface, and we shall consider only a nonpolarized interface at which at least one charged species partitions between the two phases.

4.1.1. Nonpolarized Interface

It was seen earlier for zero net current that equation (4-4) linking current and voltage simplifies to equation (4-9), which is a form of the Nernst equation—the fundamental relationship governing the operation of most potentiometric sensors. The partitioning of *only one* type of charged species across the interface is a rather unlikely situation. Usually there are several charged species which can cross the interface under the given conditions. A partial exchange current can be assigned to each of those species. The relative magnitude of those partial exchange currents can be related to the selectivity of the interface: the interface is said to be selective to the species with the highest partial exchange current.[6] According to the number of charged species involved in the overall charge transfer we divide the nonpolarized interfaces into

1. *Perm-selective,* only one type of ion can pass through.
2. *Semipermeable,* one type of ion cannot pass through.
3. *Nonselective,* all ions can pass through.

In all three cases the condition of equality of the electrochemical potential of the charged species which can cross must hold. In addition, the *condition of electroneutrality* must apply to all thick electrically conducting phases at equilibrium. In order to satisfy this condition the phase must have a minimum thickness, which is related to the dielectric constant and conductivity. Let us now assume that all the phases are sufficiently thick in order to satisfy the condition of electroneutrality. This condition is equivalent to the statement that there is no electric field (potential gradient) inside a conducting phase at zero current.

4.1.1.1. Perm-Selective Interface

This interface is the site of the *Nernst potential*, which we now derive from the thermodynamical point of view. The zero-current axis in Figure 4-1 represents the electrochemical cell at thermodynamical equilibrium, so partitioning of charged species between the two phases is described by the Gibbs equation (A-20), from which it follows that the electrochemical potential of species i in the sample phase (s) and in the electrode phase (β) must be equal,

$$\tilde{\mu}_i^\beta = \tilde{\mu}_i^s \tag{4-16}$$

The electrochemical potential of species i in a given phase contains terms originating from both the electrical and nonelectrical work associated with charge transfer across the interface,

$$\tilde{\mu}_i = \mu_i^0 + RT \ln a_i + zF\phi \tag{4-17}$$

where F is the Faraday constant, while ϕ is the electrostatic potential of the phase, the so-called Galvani or inner potential.[1] The profiles of activity, electrostatic potential, and electrochemical potential at the sample–electrode interface are shown in Figure 4-5. The equality of the electrochemical potential dictates that two terms in equation (4-17) balance out at equilibrium. It will be assumed that at time $t = 0$ there is an initial activity a_0 and that the electrostatic potential of the two phases is identical. When equilibrium is established, a small amount of cations is transferred from the sample to the electrode phase. This separation of charge results in the formation of the electrostatic potential difference between the two phases with the electrode being positive with respect to the sample. We shall return to Figure 4-5 later.

The Nernst equation, which relates the potential difference $\pi = \phi^\beta - \phi^s$ at the interface to the activities of species i in phases β and s, is obtained from equations (4-16) and (4-17) by making the two electrochemical potentials equal,

$$\pi = \pi^0 + \frac{RT}{zF} \ln \frac{a_i^s}{a_i^\beta} \tag{4-18}$$

In metal electrodes the standard state of an electron in the metal is

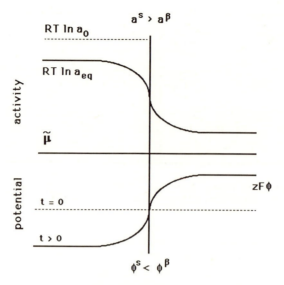

Figure 4-5. Activity–potential profiles at a nonpolarized interface. Superscripts s and β refer to the solution and solid phase, respectively. Activity at $t = 0$ is a_0.

taken to be unity. In other cases it is possible to assume that the activity inside the electrode phase is constant. For a membrane placed symmetrically between two solutions containing the primary ion, such as the case of a conventional ion-selective electrode with constant internal reference solution (S2) (Figure 4-6), the Galvani potential $\phi(\beta)$ inside the electrode phase cancels out:

$$\pi = \phi(S1) - \phi(\beta) + \phi(\beta) - \phi(S2) = RT/zF \ln a(S1)/a(S2) \qquad (4\text{-}19)$$

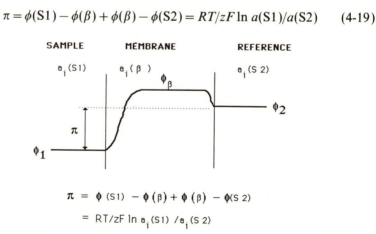

Figure 4-6. Potential profile through a perm-selective membrane placed symmetrically between the sample (S1) and a reference solution (S2).

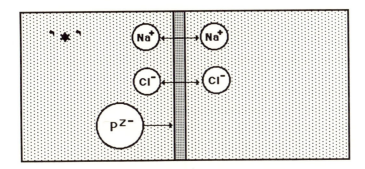

Figure 4-7. Origin of the Donnan potential. The semipermeable membrane blocks transfer of the large anion P^{z-} which is present only in the compartment ✦.

In all these cases we have a simple functional relationship between the activity of the ion in the sample and the interfacial potential,

$$\pi = \pi^0 + RT/zF \ln a_i(\text{S1}) \qquad (4\text{-}20)$$

It is important to note that the electrode potential is related to activity and not to concentration. The reason is that the partitioning equilibria are governed by the chemical (or electrochemical) potentials, which must be expressed in activities.

4.1.1.2. Semipermeable Interface

In the second case the majority of ions can pass through the interface, but some are excluded. This situation, which is depicted in Figure 4-7, is common in *dialysis* when small ions and low molecular substrates can move across the membrane but large polyelectrolyte ions cannot. This exclusion leads to the formation of the *Donnan potential* whose origin will be explained with the aid of a simple case of uni-univalent electrolyte (NaCl) and a large polyelectrolyte anion P^{z-}, which is present only in compartment "✦" and carries z negative charges. Again we invoke the two conditions which must be satisfied at equilibrium: equality of the electrochemical potential on both sides of the membrane for those species which can communicate across it, and the condition of electroneutrality which must exist in both compartments.* The first condition leads to the

---

* The equation of electroneutrality must be expressed in concentrations rather than activities. It is therefore necessary to write the Nernst equation in terms of concentrations as well and to assume that the activity coefficients on both sides of the membrane are the same and cancel out.

potential difference shown in equation (4-19) which, written for the specific ions in question, is

$$\pi = \phi(S1) - \phi(\beta) + \phi(\beta) - \phi(S2) = RT/F \ln C_{Na}^*/C_{Na} \qquad (4\text{-}21)$$

Subscript β is used for the interior of the membrane and cancels out in a symmetrical arrangement. Only one potential difference can exist across the membrane, so the partitioning of the chloride ions is uniquely related to that of the sodium ions,

$$\pi = RT/F \ln C_{Na}^*/C_{Na} = -RT/F \ln C_{Cl}^*/C_{Cl} \qquad (4\text{-}22)$$

The negative sign in the latter equation is due to the negative charge of the anion. Therefore

$$C_{Na}^*/C_{Na} = C_{Cl}/C_{Cl}^* \qquad (4\text{-}23)$$

The condition of electroneutrality dictates that

$$C_{Na} = C_{Cl} = C \qquad (4\text{-}24)$$

and

$$zC_P^* + C_{Cl}^* = C_{Na}^* \qquad (4\text{-}25)$$

Equations (4-23)–(4-25) yield, for the ion activity ratio,

$$C_{Na}^*/C_{Na} = -zC_P^*/2C + [1 + (zC_P^*/2C)^2]^{1/2} \qquad (4\text{-}26)$$

For low molar concentration of P and high concentration of NaCl, the term $(zC_P^*/2C)^2$ can be neglected in relation to 1 and equation (4-26) reduces to

$$C_{Na}^*/C_{Na} = 1 - zC_P^*/2C \qquad (4\text{-}27)$$

Thus, when the macromolecule is negatively charged $(z < 0)$ the ratio $C_{Na}^*/C_{Na} > 1$. This means that sodium ion compensates the excess of the negative charge due to the polyanion and there is a potential across the membrane given by

$$\pi = RT/F \ln(1 - zC_P^*/2C) \qquad (4\text{-}28)$$

Even for this simplest of all situations we had to make a fairly drastic assumption in order to proceed from equation (4-26) to (4-27). The situation is considerably more complicated when different multivalent ions are present in the solution, although the basic argument remains unaltered. In biological fluids, such as whole blood, the value of the Donnan potential across the dialysis membrane can be several millivolts.

4.1.1.3. Liquid Junction

A liquid junction is a solution contact which physically separates two solutions, usually the reference electrode compartment and the sample. It is not an "interface" in the usual meaning of that word, because it can take the form of a narrow channel filled with the same solvent but separating two solutions of different composition. In that channel there is a concentration gradient of all the species (including the solvent) and therefore the gradient of the chemical potential of all those species. However, because they can move from one compartment to the other without restriction, this arrangement can be viewed also as a "totally nonselective" interface. It is a site of the troublesome *liquid junction* (or *diffusion*) *potential* E_j. This potential is the result of diffusion, so it is a nonequilibrium quantity. Thus, any potentiometric measurement which uses a liquid junction is in non-equilibrium (nonthermodynamical) by definition.

The physical origin of E_j is explained in Figure 4-8. We assume that a narrow capillary separates two solutions of NaCl in which the concentrations satisfy $C_A > C_B$. Thus a concentration gradient of NaCl exists along the capillary and Na^+ and Cl^- ions move along this gradient. The mobility of Cl^- is higher ($67.35 \, ohm^{-1} \, cm^2 \, s^{-1}$) than that of Na^+ ($50.10 \, ohm^{-1} \, cm^2 \, s^{-1}$) for the same time interval, so Cl^- runs ahead of Na^+ thus causing a separation of charge. However, this creates an electric field which acts against further separation of these two ions. Thus, after some time, a steady state is reached in which the ions no longer separate and both move with the same velocity along the gradient of the chemical potential. This is called a coupled diffusion/migration. It is described by the

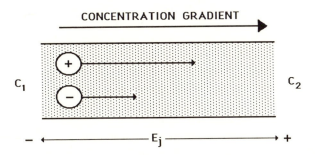

Figure 4-8. Origin of the liquid junction potential. Concentration C_1 is higher than C_2.

Nernst–Planck equation, which for the sake of simplicity is expressed in only one direction (d/dx),

$$J_i = -U_i RT \, da_i/dx - z_i FU_i a_i \, d\phi/dx \qquad (4\text{-}29)$$

where subscript i refers to species i, J is the mass flux (the amount of substance passing through unit area), U_i the absolute mobility, u_i the electrolytic mobility, and λ_i is the equivalent ionic conductivity. The two are related by

$$u_i = |z_i| \, FU_i = \lambda_i/F \qquad (4\text{-}30)$$

The diffusion [the first term in equation (4-29), because $D_i = U_i RT$] is driven by the gradient of the chemical potential which, in turn, is caused by the gradient of the activity of species i, namely da_i/dx. The potential gradient (electric field) $d\phi/dx$ is another force which acts only on ions and causes their migration [the second term in equation (4-29)].

When current passes through the electrolyte solution, the transference number t_i of ion i (the fraction of total charge transported by ion i) is expressed as

$$t_i = z_i^2 U_i a_i \Big/ \sum_j z_j^2 U_j a_j \qquad (4\text{-}31)$$

The expression for the liquid junction potential is then obtained from equations (4-29) and (4-31) by integrating over the length of the junction, L,

$$\Delta\phi = -\phi_A - \phi_B = -RT/F \int_{x=0}^{x=L} (t_i/z_i) \, d\ln a_i \qquad (4\text{-}32)$$

When the activities are replaced by concentrations, we obtain

$$\Delta\phi = \pi_j = -RT/F \int_{x=0}^{x=L} (t_i/z_i) \, d(\ln f_i + \ln C_i) \qquad (4\text{-}33)$$

Equation (4-33) cannot be integrated in general because the spatial functions of activities and transference numbers are unknown. Thus, the value of the junction potential π_j cannot be calculated except in a very special case, in which *experimental* arrangements are made to allow the introduction of some simplifying assumptions, namely, constancy of the activity coefficient, and linear concentration profiles throughout the junction. The physical implementation is a capillary (Henderson junction) separating two electrolyte solutions with a common ion,

$$\pi_j = \frac{\sum (|z_i/z_i|)\, u_i[C_i(B) - C_i(A)]}{\sum |z_i|\, u_i[C_i(B) - C_i(A)]} \frac{RT}{F} \ln \frac{\sum |z_i|\, u_i C_i(A)}{\sum |z_i|\, u_i C_i(B)} \tag{4-34}$$

In the case of a uni-univalent electrolyte, such as 0.1 M KCl//0.1 M HCl, the liquid junction potential can be calculated from the equivalent ionic conductivities λ_i of the two solutions (remember, $\lambda =$ unit conductivity/ concentration), called the *Sargent equation*,

$$\pi_j = \pm \frac{RT}{F} \ln \frac{\lambda_i(B)}{\lambda_i(A)} \tag{4-35}$$

Equation (4-35) shows explicitly the dominating influence of the ionic mobility on the value of the liquid junction potential. For the above example $\pi_j = 28.52$ mV. If the HCl solution is replaced by the NaCl solution with the same concentration, $\pi_j = 4.86$ mV.

Unfortunately, in practice, combination of the solutions on either side of the junction cannot be chosen at will. For aqueous solutions a good practical compromise is the use of a *salt bridge*, which is a combination of two liquid junctions connected back-to-back (Figure 4-9). The internal solution is chosen such that both junction potentials are *dominated* by the salt possessing identical mobility of the anion and cation. An example is a saturated aqueous solution of KCl ($U_{K^+} = 7.619 \times 10^{-4}$ cm^2 s^{-1} V^{-1}, $U_{Cl^-} = 7.912 \times 10^{-4}$ cm^2 s^{-1} V^{-1}). The internal salt-bridge solution can leak out and contaminate the sample, so it is desirable to have other salts available which have approximately equal mobilities. These are, for example, Li acetate (4.010 and 4.24) and NH_4NO_3 (7.61 and 7.404). By using a salt bridge the liquid junction potential can be reduced to less than 1 mV.

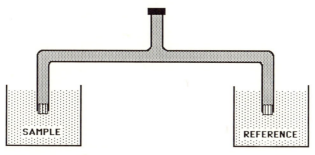

Figure 4-9. Diagram of a salt bridge. The internal filing solution contains a salt possessing identical mobility of the cation and anion.

By far the biggest problems with the stability and the magnitude of the liquid junction potentials arise in applications where the osmotic or hydrostatic pressure, temperature, and/or solvents are different on either side of the junction. For this reason the use of an aqueous reference electrode in nonaqueous samples should be avoided at all cost, because the gradient of the chemical potential of water has a very strong effect on the activity coefficient gradients of the ions in equation (4-32). In order to circumvent these problems one can use a junction containing the same solvent as the sample; the internal reference compartment, or *internal reference ion method*, usually works quite well. In this case we use an electrode in a compartment with constant activity of the primary ion in a solution whose composition is as close as possible to that of the sample.

4.1.1.4. Mixed Potential

The last type of potential sometimes encountered at the nonpolarized interface, and which needs to be discussed, is called a *mixed potential*. Despite the fact that the net current flowing through the interface is zero, this is *not an equilibrium potential*. It can be found at membranes whose partial exchange current densities are low and are all of the same order of magnitude, i.e., at ion-selective membranes possessing very low selectivity. The feature distinguishing the mixed potential from the liquid junction potential (which is also nonselective) is the *magnitude of the sum of the partial current densities*. For a mixed potential it is typically less than 10^{-6} A cm^{-2}. For a physical explanation of the origin of this potential, reference must be made to the current–voltage equation (4-4), which can be written for any charged species that crosses the interface, including ions. If two different ions are crossing the interface between the membrane and the solution, the partial exchange current density can be obtained for each of them from the current–voltage equation. At zero current two conditions apply: (1) there is only one potential at the interface common to all ions, and (2) the sum of all partial currents must be zero.

Even for uni-univalent electrolytes, the solution of these equations leads to intractable algebra and some drastic assumptions are required in order to obtain an explicit solution.[6] Figure 4-10 presents graphically the origin of the mixed potential. Movement of the ions to and from the membrane is shown as superscript arrows. The equilibrium potentials and exchange-current densities for each individual process are different, so the mixed potential E_{mix} will lie somewhere between the two equilibrium potentials and hence satisfy the condition of zero net current. If the rate of the individual charge-transfer reaction changes, e.g., due to some specific adsorption or relative motion of the electrolyte with respect to the stationary membrane (stirring), then the position of E_{mix} will also change.

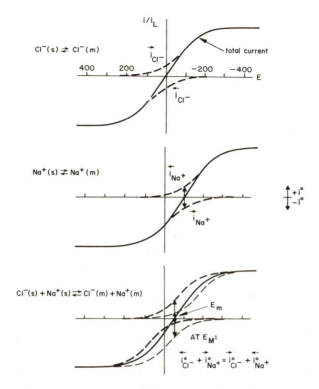

Figure 4-10. Origin of the mixed potential. The arrows above partial currents correspond to the equilibria shown on the left. The polarity of the current shown on the right corresponds to the solution.

It is important to realize that different ions, and their corresponding exchange-current densities, respond differently to these effects. This can explain why a seemingly selective response is obtained for certain kinds of adsorption, such as immunochemically mediated adsorption.

There is an interesting conceptual connection between the mixed potential and the true equilibrium potential. If the exchange-current density for one ion is very high relative to the other partial exchange current densities, then the value of E_m will lie close to the equilibrium potential of this *dominating* charge-transfer process and the membrane will behave as a true equilibrium sensor, the equilibrium potential being set up by the dominating ion. Despite the fact that the complicated algebra so far prevents the potentiometric selectivity coefficients being expressed in terms of the partial exchange current densities, this is a very attractive concept because it allows at least a phenomenological interpretation of the potentiometric behavior of low-exchange current density membrane electrodes.

Table 4-1. Estimates of Apparent Standard Exchange Current Densities i_0, Exchange Current Ratios $i_{0,I}/i_{0,K}$, and Selectivity Coefficients for a K^+ ISE Membrane[a]

Solution	i_0 (A cm^{-2})	$i_{0,I}/i_{0,K}$	$K_{K/I}$
KCl	2.1×10^{-3}	1	1
RbCl	4.7×10^{-3}	3.2	2.4
CsCl	5.3×10^{-4}	0.32	0.34
NH$_4$Cl	4.2×10^{-5}	4×10^{-2}	2×10^{-2}
NaCl	4.2×10^{-6}	1×10^{-3}	2×10^{-4}
LiCl	2.6×10^{-6}	7×10^{-4}	5.3×10^{-5}
N(CH$_3$)$_4$Cl	2.1×10^{-6}	6.6×10^{-4}	4.6×10^{-5}

[a] The membrane phase contains 2.7×10^{-3} M valinomycin in diphenyl ether; the geometrical surface is about 0.012 cm^2. The concentration of the interfering ion is 0.1 M. (After Cammann[6].)

As expected, the exchange-current densities and selectivity coefficients are closely related, as is evident from the data obtained experimentally for a K^+ selective membrane (Table 4-1).

4.1.1.5. Reference Potential

The need for a rather lengthy discussion of the liquid junction potential was prompted by the fact that, in potentiometric measurement, the analytical information is obtained by measuring the cell voltage E_{cell} given by

$$E_{cell} = \pi_{ind} - (\pi_{aux} + |\pi_j|) \qquad (4\text{-}36)$$

Fifty percent of this information comes from the indicator electrode and the rest from the reference system, including the liquid junction. The fact is that the electrical circuit must be completed and that the probability of failure is equally distributed between the three terms in equation (4-36). For amperometric measurements, the auxiliary electrode involves no special requirement except that it be large as compared to the indicator electrode. This argument becomes evident when we examine curve B in Figure 4-1 and realize that what is plotted is *current density*. In that case the shape of the *i–E* curve depends only on processes taking place at the indicator electrode which, because it is smaller, possesses a higher current density. When the cell is used in the limiting-current region, the potential of the *auxiliary electrode* will be close to the equilibrium potential because the current density at that electrode is very small. The *i–E* curve is steep in the proximity of E_{eq}, so the potential of the auxiliary electrode will be virtually constant. In older electrochemical literature many amperometric measurements were made and referenced against the potential of a large auxiliary electrode. Obviously, for zero-current potentiometry the argument based on the relative size of the electrodes is not valid and we must look for a different kind of reference potential.

There are three main requirements which such a potential must satisfy. It must be stable, reversible, and reproducible. *Stable* means that it will not change when the composition of the sample changes. *Reversible* means that it will return rapidly to its equilibrium value after a small transient perturbation. This condition implies a low charge-transfer resistance R_{ct} (high exchange-current density). *Reproducibility* implies that the same electrode potential will always be obtained when the reference electrode is constructed from the same electrode/solution combination. These requirements are common to all potentiometric sensors.

The requirement of reversibility (good ohmic contact) is easily satisfied by selecting an electrochemical reaction which is very fast, i.e., has a very high exchange-current density ($> 10^{-3}$ A cm^{-2}). An example of such a reaction is the silver/silver chloride electrode, which is a silver metal coated with a few microns of silver chloride. Reference electrodes utilizing an insoluble salt are called *reference electrodes of the second kind.* For this electrode the primary electrochemical reaction is

$$Ag^+ + e \Leftrightarrow Ag \tag{4-37}$$

Thus

$$\pi = \pi^0_{Ag} + RT/F \ln a_{Ag}, \qquad \pi^0_{Ag} = +0.799 \text{ V}$$

AgCl is a relatively insoluble compound, hence its purpose is to maintain the activity of silver ions at the electrode constant,

$$AgCl \overset{K_{sp}}{\Longleftrightarrow} Ag^+ + Cl^- \tag{4-38}$$

$$K_{sp} = a_{Ag} a_{Cl} = 10^{-10.2} \text{ at } 25 \text{ °C}$$

Combination of equations (4-37) and (4-38) gives the potential of the silver/silver chloride electrode,

$$\pi = \pi_{AgCl^0} - RT/F \ln a_{Cl}, \qquad \pi_{AgCl^0} = +0.2223 \text{ V} \tag{4-39}$$

This potential is a function of the activity of chloride ions; the latter quantity must be kept constant in order to satisfy the requirement of stability of this potential. Thus, the Ag/AgCl electrode must be stored in a separate compartment with defined and constant activity of chloride ion. However, the electrical circuit must be completed, therefore this compartment must be in electrical contact with the sample in which the indicator electrode is placed. This contact is realized by the liquid junction, which adds its own diffusion potential to the overall cell voltage. Obviously, it is not possible to use an electronic conductor as the junction because it would create its own interface with the sample.

A schematic diagram of a silver/silver chloride electrode is shown in Figure 4-11. It is customary to write an electrochemical cell with interfaces denoted by / and junctions by //. For a simple cell consisting of a reference electrode and an indicator electrode, we therefore write

$$Ag/AgCl/ \text{ (c) } KCl// \text{ sample /indicator electrode}$$

Thus a complete reference electrode consists of three parts: the internal reference element (Ag/AgCl), a reference solution compartment [(c) KCl], and a liquid junction //. The volume of the reference solution compartment in a practical laboratory reference electrode is usually 2–5 ml. The liquid junction is an open channel, so there is an outflow of the internal reference solution equal to approximately 2 nl h^{-1} (Figure 4-12). This flow actually helps to stabilize the liquid junction potential.[7] The lifetime of such a reference electrode depends on the conditions under which it is used, but it often exceeds one month of continuous use. The condition of reproducibility for a fast electrode reaction is almost automatic.

Other reference electrodes are also used. One example is a calomel electrode which can be represented as

$$Hg/Hg_2Cl_2/(c) \text{ } KCl// \text{ sample...}$$

or a silver electrode which consists of silver wire immersed in a (usually nonaqueous) solution of Ag$^+$. In fact any ion-selective electrode discussed below in Section 4.1.2 can be used as a reference, provided that it is kept in a solution possessing constant activity of the primary ion.

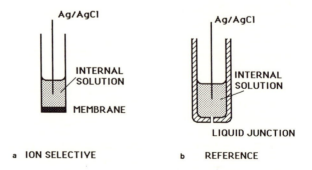

Figure 4-11. Diagram of (a) an ion-selective electrode and (b) a reference electrode.

Figure 4-12. Photographs of various designs of glass–liquid junctions. The dark streak is a dye which helps to visualize the outflow of the internal solution from the junction. (The photographs were kindly provided by W. Simon and are reprinted from Ref. 7 with permission of the American Chemical Society.)

In order to avoid confusion when reporting electrochemical data, hydrogen electrode (Figure 4-13) has been chosen arbitrarily as the primary reference electrode. By convention the *standard potential* of the hydrogen electrode, represented schematically as

$$Pt/Pt \text{ black}, H_2(atm), 1N \, H_2SO_4/...$$

is set to zero ($\pi^0_{H/H^+} = 0$), thus establishing a hydrogen scale of standard potentials. This electrode is impractical and *secondary reference electrodes*, such as those discussed above, are used instead.

There have been many attempts to prepare a microreference electrode which would be comparable in size to the integrated ion sensors, such as ion-sensitive field-effect transistors (ISFET) (Section 4.1.3.2.3). These attempts follow broadly three lines of approach: scaling down of a macroscopic reference electrode,[8,9] elimination of the reference solution compartment while preserving the internal element structure (e.g., Ag/AgCl), and utilization of "inert" materials such as polyfluorinated hydrocarbons, particularly in the so-called "reference FET" configuration. The last type clearly violates the requirement of high exchange-current density: there is always some potential at any interface. Unless that potential is dominated by a high exchange-current density reaction, it cannot be stable. The second type (internal element only) violates the condition of constant activity of the reference ion (e.g., a_{Cl}). Although such an electrode will form a good ohmic contact with the solution, its potential will change with the composition of the solution. The first type, a miniature conventional reference electrode, can function, but only for a short period of time,

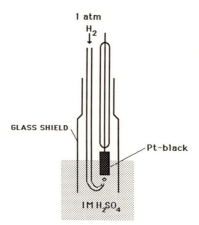

Figure 4-13. Schematic diagram of a normal hydrogen electrode. A "standard" hydrogen electrode is defined for "unit activity of the hydrogen ion" and cannot be realized exactly.

depending on the actual volume of the reference solution compartment and the conditions under which it is used. The usable lifetime is only minutes to hours. The fact that such electrodes cannot sustain an outflow of the internal solution makes the liquid junction potential unstable in practical applications. In short, there is no truly "microreference electrode" available.

4.1.2. Ion-Selective Membrane

An ion-selective membrane is the key component of all potentiometric ion sensors. It establishes the preference with which the sensor responds to the determinand (ion of interest) in the presence of various other ionic components of the sample. By definition, the ion-selective membrane forms a nonpolarized interface with the solution. If the interface is permeable to only one ion it is called "perm-selective" (Section 4.1.1.1), and the potential difference at that interface is expressed by the Nernst equation (4-18). If more than one ion can permeate, the interface can lie anywhere between the liquid junction (Section 4.1.1.3) and the mixed potential (Section 4.1.1.4). The key distinguishing feature between the two is the *absolute magnitude* of the total exchange-current density, this being high for the former and low for the latter.

On the other hand, the notion of ion selectivity depends on the *relative magnitude* of the exchange-current density of one ion (determinand x) relative to the other ions (interferants i). The higher this ratio the more selective the membrane. A well-behaved membrane, i.e., one which is stable, reproducible, immune to adsorption and stirring effects, and also selective, has high both absolute and relative exchange-current density. However, a poorly behaving membrane with low exchange-current density may still show some selectivity but it will be affected by all the above factors. Although this formulation of the selectivity for ion sensors is conceptually correct so far, it has not led to an explicit expression for, or explanation of, the physical meaning of the selectivity constant $K_{x,i}$ in the *Nikolskij–Eisenman* equation

$$E_{out} = RT/z_x F \left[\ln \left(a_x + \sum_i K_{x,i} a_i^{z_x/z_i} \right) \right] \tag{4-40}$$

It can be seen from this equation that the membrane is more selective for the determinand x the lower the value of each selectivity constant $K_{x,i}$. Thus, our ability to manipulate the components of $K_{x,j}$, which affect the value of this constant, and our ability to design more selective membranes depend on the identification of the mechanistic details that constitute the selectivity constant. In this section we shall discuss the individual processes affecting the membrane selectivity. A detailed treatment of this subject can be found in specialized books.[10–12]

According to the nature of the binding sites, the membranes can be divided into those containing fixed ionic sites, mobile ion-exchanger, or membranes containing neutral ionophores. The binding sites are incorporated in the membrane matrix, which controls the internal dielectric constant, lipophilicity, transport, and other mechanical and electrical properties of the membrane. Although ion-selective membranes are usually discussed in the context of sensing in an aqueous environment, water is not the only usable medium in which these sensors can operate. There have been many successful applications of ion-selective electrodes (ISE) in non-aqueous media and at high temperatures and pressures. The suitability of any given sensor for such applications is usually dictated by its construction and by the material aspects, rather than by some fundamental restrictions on the operational principles of the device.

4.1.2.1. Membranes with Fixed Ionic Sites

The best-known example of this type of membrane is the glass electrode in which the anionic sites are created by defects in the SiO_2 matrix, and by the cationic vacancies due to the nonsilicon constituents of the glass.[13] When the glass membrane is exposed to water, an approximately 50–1000 Å thick hydrated layer is formed. The chemical composition of the glass in this layer is the same as that in the dry bulk, however, the mobility of the ions is much lower than in water ($D \sim 10^{-11}\,cm^2\,s^{-1}$). The concentration of anionic binding sites is estimated between 3 and 10 M, which accounts for the wide dynamic range of these sensors. The membrane is usually blown into a bulb of typical wall thickness 50–200 μm. The optimum thickness of the wall is the result of compromise between the mechanical stability and electrical resistance. The latter is usually on the order of 10 MΩ. There are two processes which take place during interaction of the hydrated layer and sample: ion exchange and diffusion of all participating ions. We shall see that they both contribute to the value of the selectivity coefficient.

A simple case is now considered of only two univalent ions Na^+ and H^+ exchanging between the solution and membrane. Therefore, their electrochemical potentials in these two phases must be equal: $\tilde{\mu}_H^s = \tilde{\mu}_H^m$ and $\tilde{\mu}_{Na}^s = \tilde{\mu}_{Na}^m$. The ion exchange takes place according to the equation

$$H^+(m) + Na^+(s) \overset{K_{ex}}{=\!=\!=} Na^+(m) + H^+(s) \qquad (4\text{-}41)$$

where the equilibrium ion-exchange constant K_{ex} for this process is given by

$$K_{ex} = a_{Na}^m a_H^s / a_H^m a_{Na}^s \qquad (4\text{-}42)$$

s and m referring to the solution and membrane, respectively. The electrochemical potentials can be expanded [see equation (4-17)] and rearranged to yield the expression for the Donnan potential (Section 4.1.1.2),

$$\phi^m - \phi^s = \frac{\mu_H^{0,s} - \mu_H^{0,m}}{F} + \frac{RT}{F} \ln \frac{a_H^s}{a_H^m} \qquad (4\text{-}43)$$

It should be noted that in the interest of generality we have not specified in which arrangement the membrane will be used. Here the potential is examined at the sample–membrane interface, but the same relationship would apply also for the second interface between the membrane and internal reference electrode compartment in a symmetrical arrangement (Section 4.1.3.1). According to the model shown in Figure 4-14 we must add the second contribution to the membrane potential, that from the diffusion potential. In the simplest form it is obtained from the Henderson equation (4-34) with all the limitations possessed by this equation. The charge transport in the dry part of the membrane (m′) is conducted only by sodium ions, so their mobility of the hydronium ion is taken to be zero ($u_H^{m'} = 0$). Both ions are univalent, as a result of which equation (4-34) further simplifies to

$$\phi^{m'} - \phi^m = \frac{RT}{F} \ln \frac{u_H a_H^m + u_{Na} a_{Na}^m}{u'_{Na} a_H^{m'}} \qquad (4\text{-}44)$$

The potential difference between the interior of the solution and the interior of the membrane is obtained by adding equations (4-43) and (4-44),

$$\phi^{m'} - \phi^s = \text{const} + \frac{RT}{F} \ln \frac{(u_{Na}/u_H)\, a_{Na}^m a_H^s + a_H^m a_H^s}{a_H^m} \qquad (4\text{-}45)$$

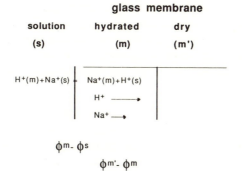

glass membrane

solution	hydrated	dry
(s)	(m)	(m')

H⁺(m)+Na⁺(s)   Na⁺(m)+H⁺(s)

H⁺ ──────▸

Na⁺ ──────▸

ϕ^m- ϕ^s

$\phi^{m'}$- ϕ^m

Figure 4-14. Development of the potential in the membrane of a glass electrode.

where the constant includes the difference between the standard chemical potentials of the hydrogen ion in the hydrated layer and in the solution. In the symmetrical arrangement of the membrane (Section 4.1.3.1) this term cancels out. The ratio of the mobility of Na^+ in the dry membrane to the mobility of the hydrogen ion in the hydrated part, and the activity of the Na^+ in the dry part, are also included:

$$\text{const} = \frac{\mu_H^{0,s} - \mu_H^{0,m}}{F} + \frac{RT}{F}\ln(u'_{Na}/u_H)\, a_{Na}^m \tag{4-46}$$

Equation (4-42) for the ion-exchange equilibrium is next substituted into equation (4-45),

$$\phi^{m'} - \phi^s = \text{const} + \frac{RT}{F}\ln[a_H^s + (u_{Na}/u_H)\, K_{ex}\, a_{Na}^s \tag{4-47}$$

The potentiometric selectivity coefficient $K_{x,i}$ is then given by

$$K_{x,i} = (u_{Na}/u_H)\, K_{ex} \tag{4-48}$$

This equation shows that in order to obtain a selective membrane, the sodium ion mobility in the hydrated layer must be small relative to that of the hydrogen ion while the value of the ion-exchange constant must also be small. Expansion of the selectivity coefficient to include selectivity to other ions would involve inclusion of more complex ion-exchange equilibria and a more complex form of the Henderson equation. This rapidly leads to intractable algebra, nevertheless the concept of the physical origin of the selectivity coefficient remains the same.

Other types of membrane with fixed sites include single crystals of sparsely soluble salt and heterogeneous membranes in which the insoluble salt is imbedded in some suitable inert binder. Obviously, in order to consider these membranes at equilibrium it is necessary in principle to use saturated solution. In practice, these membranes are often used in solutions which are below saturation. In that case the "insoluble salt" slowly dissolves.

An example of such an electrode is the Ag/AgCl electrode discussed in Section 4.1.1.5. Although the primary charge-transfer reaction at this interface is that of the silver ion [equation (4-37)] the electrode will respond also to the chloride ion [equation (4-38)] owing to the solubility product of AgCl. We now examine what happens if other ions, which also form insoluble silver salt, are present in the solution. If the solubility product of the other salt AgX is lower than that of silver chloride and the activity a_X is sufficiently high,

$$K_{sp}(AgCl)/K_{sp}(AgX) > a_{Cl}/a_X \qquad (4\text{-}49)$$

the electrode will undergo metathesis and its surface will become partially covered with the solid phase of AgX. Clearly, this is a nonequilibrium situation which reflects the nonequilibrium nature of the Nikolskij–Eisenman equation. Interference of this kind is particularly severe if sulfide ion is present, because sulfides have exceptionally low solubility products.

On the other hand, the low value of the solubility product of silver sulfide can be used to advantage. It is possible to prepare membranes in which Ag_2S or HgS are used as a matrix and a more soluble salt is present as an additive. Examples of such membranes are given in Table 4-2. Let us consider a mixed PbS–Ag_2S membrane for which the respective solubility products are given as

$$K_{sp}(Ag_2S) = a_{Ag}^2 a_S = 6 \times 10^{-50} \qquad (4\text{-}50)$$

and, at 25 °C,

$$K_{sp}(PbS) = a_{Pb} a_S = 2.5 \times 10^{-27}$$

The "equilibrium" activity of lead ion is obtained from equation (4-50) in the form

$$a_{Pb} = [K_{sp}(PbS)/K_{sp}(Ag_2S)] \, a_{Ag}^2 \qquad (4\text{-}51)$$

Table 4-2. Membranes for Solid State Electrodes

Primary ion	Orion electrode[a]	Homogeneous membranes[b]	Heterogeneous membranes[c]
F^-	LaF_3	—	—
Cl^-	$AgCl/Ag_2S$	Hg_2Cl_2/HgS, $AgCl$	$AgCl$
Br^-	$AgBr/Ag_2S$	Hg_2Br_2/HgS, $AgBr$	$AgBr$
I^-			
CN^-	AgI/Ag_2S	AgI	AgI
Hg^{2+}			
SCN^-	$AgSCN/Ag_2S$	$Hg_2(SCN)_2/HgS$, $AgSCN$	$AgSCN$
S^{2-}	Ag_2S	Ag_2S	Ag_2S
Cu^{2+}	Cu_xS/Ag_2S	$CuSe$	Cu_xS, Cu_xS/Ag_2S
Pb^{2+}	PbS/Ag_2S		PbS/Ag_2S
Cd^{2+}	CdS/Ag_2S		CdS/Ag_2S

[a] Typical commercial electrode.
[b] Pressed pellet or a single crystal.
[c] Solid powder in an organic matrix.

This would be the activity of lead present in the solution due to dissolution of the membrane. It therefore represents the detection limit. The "detectable" a_{Pb} is added to this background value. Silver ion is the charge-transferring species, hence the membrane potential is expressed in terms of the silver-ion activity which is obtained from equation (4-50). Thus

$$\phi^m = \phi_{Ag/Ag^2S} + \frac{RT}{2F} \ln \frac{K_{SP}(Ag_2S)}{K_{SP}(PbS)} + \frac{RT}{2F} \ln(a_{Pb} + K_{Pb/Ag}a_{Ag}) \quad (4\text{-}52)$$

Moreover, since Ag^+ is the only ion which penetrates into the interior of the solid phase, the relative mobility is not a factor in the selectivity coefficient, which is expressed only as the ratio of the solubility products of the two salts,

$$K_{Pb/Ag} = K_{SP}(Ag_2S)/K_{SP}(PbS) \quad (4\text{-}53)$$

The fact that metathesis takes place means that equations (4-52) and (4-53) have only a semiquantitative meaning.

There are only a few analytical methods that allow simple selective determination of fluoride ion. For this reason the LaF_3 single-crystal membrane ion-selective electrode is one of the most important sensors of this type. It is based on the equilibrium

$$LaF_3 = La^{3+} + 3F^-$$

This electrode is also important from the historical viewpoint because it is generally regarded as the first modern ion-selective electrode[14] which started the renaissance of zero-current potentiometry that continues to the present time.

Single-crystal lanthanum fluoride is a wide-bandgap semiconductor in which the electrical conductivity is due only to the mobility of fluoride ions. It will be shown later (Section 4.1.3) that consideration of the ionic and/or electronic conductivity of the membrane plays a critical role in the design of the internal contact in nonsymmetrical potentiometric sensors. It is also noteworthy that the fluoride electrode does not respond to the La^{3+} ion owing to its slow ion exchange.

4.1.2.2. Liquid Membranes

The title of this section is a bit of a misnomer, because these membranes are usually "solid," mainly for practical reasons. However, the mobile binding "sites" are complexing agents (ionophores), which are

Table 4-3. Examples of Electrodes Based on Liquid Ion Exchangers[a]

Ion	Form of membrane	Active material	Solvent mediator
Ca^{2+}	Liquid	Calcium di-(n-decyl)phosphate	Di-(n-octylphenyl)phosphonate
	Solid (PVC)	Calcium di-(n-decyl)phosphate	Di-(n-octylphenyl)phosphonate
NO_3^-	Solid (PVC)	Tris(substituted 1,10-phenanthroline)nickel(II) nitrate	p-Nitrocymene
ClO_4^-	Liquid	Tris(substituted 1,10-phenanthroline)iron(II) perchlorate	p-Nitrocymene
BF_4^-	Liquid	Tris(substituted 1,10-phenanthroline)nickel(II) tetrafluoroborate	p-Nitrocymene
Divalent cations ("water hardness")	Liquid	Calcium di-(n-decyl)phosphate	Decanol
U(VI)	Solid (PVC)	Di(2-ethylhexyl) phosphoric acid	Diamylamyl phosphonate
Cl^-	Liquid	Dimethyl-dioctadecylammonium chloride	

[a] Reprinted from Ref. 12 with permission of Elsevier.

dissolved in a suitable solvent and usually trapped in a matrix of organic gel. Ion activity measurements are conducted predominantly in aqueous media, so all the membrane constituents are hydrophobic. Therefore, the primary interaction between the ion in water and the hydrophobic membrane containing the ionophore is the *extraction* process. In principle, membranes of this type would function also in organic solvents and a few reports of such applications have been published. However, the major limitation is the solubility of the membrane itself and/or the extraction of the ionophore or of the solvent from the membrane. The most popular matrix for the solvent and for the ionophore is high-molecular-weight poly(vinyl chloride), which constitutes typically 30% (by weight) of the membrane. The remaining major component is the solvent (plasticizer), which ensures the mobility of the free and complexed ionophore, sets the dielectric constant, and provides suitable mechanical properties of the membrane. The ionophore is usually present in 1% amount, which corresponds to a concentration of binding sites of approximately 10^{-2} M. This is relatively low as compared to the glass electrode. Other membrane components include large hydrophobic anions, such as tetrafluoroborate, and polysiloxane. They both extend the upper limit of the dynamic range by minimizing the Donnan exclusion failure,[15] while polysiloxane improves adhesion of the membrane to the insulator in the ISFET.[16]

There are two kinds of ionophores: charged ones, which are usually termed *liquid-exchangers* (Table 4-3), and *neutral carriers* (Table 4-4). They are mobile in both the free and complexed forms, so the mobilities of all species are part of the selectivity coefficient together with the ion-exchange equilibrium constant. The best-known neutral carrier ionophore is valinomycin (Figure 4-15) which exhibits 1000:1 selectivity for K^+ in preference to Na^+ and no pH dependence. An example of a selective ion-exchanger is dioctyl phenylphosphate (Figure 4-16).

Table 4-4. Examples of Neutral Carrier-Based Electrodes[a]

Ion	Form of membrane	Active material	Solvent mediator
K^+	Solid (PVC)	Valinomycin	Dioctyladipate
	Solid (silicone rubber)	Valinomycin	—
NH_4^+	Liquid	Nonactin/monactin	Tris(2-ethylhexyl) phosphate
Ca^{2+}	Solid (PVC)	ETH 1001	o-Nitrophenyl octyl ether
Ba^{2+}	Liquid	Nonylphenoxy poly(ethylene oxy) thanol Ba^{2+} (tera phenylborate)	p-Nitroethylbenzene

[a] Reprinted from Ref. 12 with permission of Elsevier.

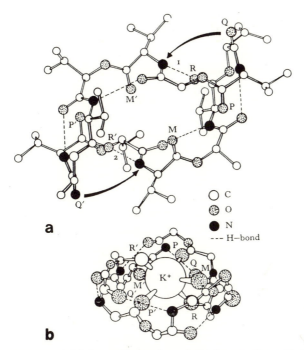

Figure 4-15. Structure of (a) free valinomycin and (b) its complex with K$^+$. Note the cooperation of the inner oxygen atoms P, M, and R in the complexation process. (Reprinted from Ref. 11 with permission of Cambridge University Press.)

Another important consideration, particularly from the standpoint of their use in solid-state sensors, is the optimum thickness of the membrane. For the membrane interior to be electrically neutral, the total thickness must be greater than the combined thicknesses of the space charges which extend on both sides of the membrane.[12] The thickness of the space charge $1/\kappa$ depends on the dielectric constant ε of the membrane and on the valency and concentration of the binding sites,

$$1/\kappa = \left(\varepsilon RT/z^2F^2 \sum_i C_i \right)^{1/2} \qquad (4\text{-}54)$$

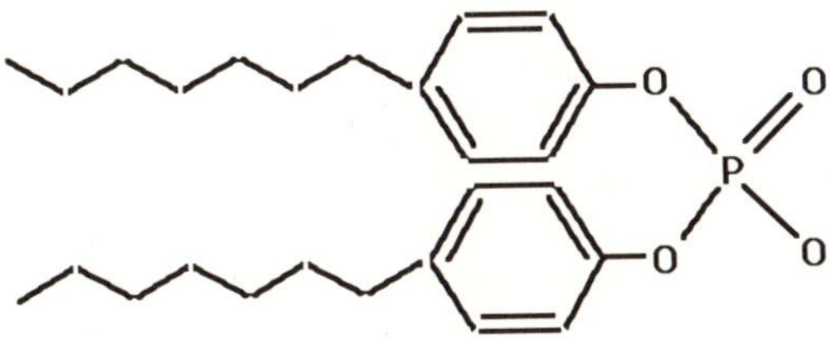

Figure 4-16. Example of a liquid ion-exchanger: dioctyl phenyl phosphate.

It is typically on the order of a thousand angstroms. In reality the minimum thickness for polymeric membranes is 50 μm or above, far more than would be expected from equation (4-54). This is probably due to the fact that these membranes hydrate to some extent and develop electrical shunts. This is usually not a problem with macroscopic ISE, which use relatively thick membranes (about 500 μm). On the other hand, in the construction of solid-state potentiometric ion sensors it is desirable to have as thin membranes as possible in order to make their application compatible with common integrated-circuit fabrication technology.

The time response $\Delta\tau$ of the membrane potential to the step variation in ion activity from $a_i(1)$ to $a_i(2)$ is

$$\Delta\tau = \frac{RT}{F}\ln\left\{1 - \left[1 - \frac{a_i(1)}{a_i(2)}\right]\exp(-t/\tau)\right\} \qquad (4\text{-}55)$$

The value of the time constant ranges from 30 μs for an AgBr membrane to 200 ms for a pH glass electrode. Often quoted response times of ISE in seconds are usually due to the time constants of associated processes, such as diffusion through the stagnant layer in the solution, and/or due to the time response of the electronics.

4.1.3. Electrochemical Cell

We now return briefly to Figure 4-1 and examine the nonpolarized interface within the context of the whole electrochemical cell. That interface had the electrode being charged positively with respect to the solution due to the transfer of cations from the solution to the electrode phase. Now we have two phases which are in electrochemical equilibrium with respect to each other, but within each phase the condition of electroneutrality is not satisfied. What happens when we now wish to use this interface in order to obtain analytical information about the composition of the solution using equation (4-20)? An electrochemical cell must be constructed and connected to an electrometer with high input resistance, so that no current can pass through it. Such an experiment is now set up and it is assumed that the indicator interface is silver metal in contact with solution containing soluble silver salt, such as AgF. In the first instance we shall complete the circuit with a lanthanum fluoride electrode which, as we already know, is reversible to fluoride ions. The internal contact of this electrode will be made of AgF (solid) and Ag. The external circuit will consist of silver wire and an electrometer in which the cell voltage E_{cell} will be measured (Figure 4-17). Thus the cell can be represented as

$$\text{Ag/AgF(solid)/LaF}_3\text{(solid)/(a) Ag}^+ \text{F}^-; 1 \text{ M KF/Ag}$$

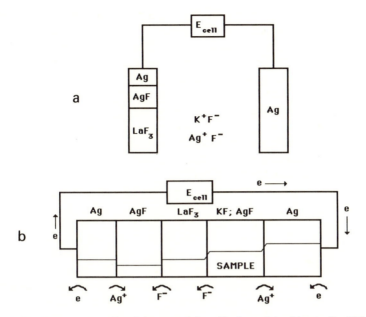

Figure 4-17. Diagram of a cell and the potential profile through it without a liquid junction.

The 1 M KF is added to the sample solution in order to ensure electrical conductivity. Although seemingly bizzare, such an electrochemical cell can be constructed and will function. When a few moles of AgF are added to the solution and the cell attains equilibrium, the resulting processes at each of the interfaces are shown in Figure 4-17 by arrows. Electrons supplied through the external silver wire will ensure charge neutrality inside the Ag layer while neutralization of the ionic charge occurs in AgF (solid). The same amount of F^- is added to and taken away from the LaF_3, so that there is no excess charge in that phase. Charge separation will take place at both LaF_3/sample and sample/Ag interfaces, so the overall change in E_{cell} will be twice that of each individual interface. Yet, charge neutrality in the solution will be preserved and this cell will measure the activity of *both* the silver *and* fluoride ion, namely, the activity of silver fluoride

$$a_F \times a_{Ag} = a_{AgF} \tag{4-56}$$

It should be noted that there are no extra thermodynamical assumptions involved, but the cell measures the *activity of the salt* and not of the single ion. Let us now rearrange the cell and use a reference electrode with a liquid junction (Figure 4-18). In this case we choose the internal reference solution which consists of 1 M KNO_3 and 0.001 M $AgNO_3$; the sample

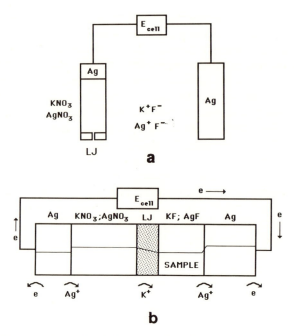

Figure 4-18. Diagram of a cell and the potential profile through it with a liquid junction.

is again 1 M KF and a small amount of AgF. The cell can now be represented as

$$\text{Ag}/1 \text{ M KNO}_3, 1 \text{ m MAgNO}_3//(a)\text{Ag}^+\text{F}^-; 1 \text{ M KF}/\text{Ag}$$

The equilibration will cause the flow of electrons in the external circuit as before, causing an equivalent amount of Ag^+ to be dissolved into the reference solution compartment. The condition of electroneutrality must be maintained, so the equivalent amount of *cations* will leak through the liquid junction to the sample compartment (in the ratio 1 Ag^+ : 1000 K^+). The actual number of ions required to establish the potential difference at the interface can be estimated. A typical value of the double-layer capacitor is 10^{-5} F cm^{-2}. If a 100 mV potential difference is to be established, this capacitor must be charged by $Q = C_{dl}\pi = 10^{-6}$ coulomb. From Faraday's law [equation (4-3)] we see that it corresponds to approximately 10^{-11} mole cm^{-2} or 10^{12} ions per square centimeter of electrode surface. Thus, a finite amount of potential-determining ions is removed from the sample. This is not really surprising, because we are dealing with a partitioning process.

The next question to be examined is where this potential difference appears in the measuring circuit. The electrometer in Figures 4-17 and 4-18 is a capacitor, because it must present an infinitely high resistance to DC current. With the electron flow as shown in the two figures the voltage appears across this capacitor, which in a modern electrometer would be a field-effect transistor. If another capacitor is placed in series with the electrometer capacitor, the voltage is divided according to the reciprocal values of these two capacitors. If the serial capacitor is undefined and variable in time (such as an iterruption in the circuit or an electrically blocking interface), a meaningful reading of the interfacial potential cannot be obtained. This is another way of realizing the need to close the electrical circuit in potentiometric measurements. It can be stated as a one-capacitor rule: *In a potentiometric measurement only one capacitor in series with the measured interface is allowed.* This also necessitates analysis of the entire measuring circuit and not only of the ion-selective electrode alone.

During the concentration step the potential profile through the reference solution compartment and through the liquid junction will be (approximately) unchanged. The word "approximately" in the preceding

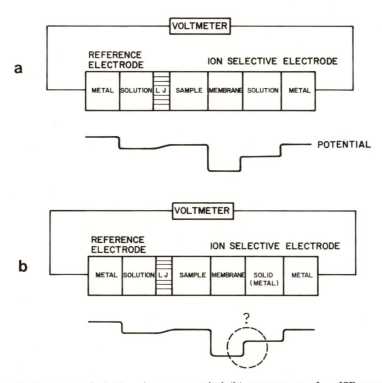

Figure 4-19. Symmetrical (a) and nonsymmetrical (b) arrangement of an ISE membrane. The question mark (?) emphasizes the critical interface.

sentence summarizes the thermodynamical uncertainity associated with the liquid junction. As noted previously, it is a nonequilibrium element because it is the site of the diffusion process. Nevertheless, within the limitation of this uncertainty it allows the potential of the indicator electrode to be measured (approximately). Thus, in summary, the use of a liquid junction provides a *practical* possibility of measuring the potential of the indicator electrode at the cost of a relaxation in thermodynamical rigor. In contrast, a thermodynamically correct measurement with the cell shown in Figure 4-17 yields only the activity of the whole salt.

The ion-selective membrane can be used in two basic configurations (Figure 4-19). If the solution is placed on either side of the membrane, such an arrangement is termed *symmetrical*. It is found in conventional ion-selective electrodes in which the internal electrical contact is provided through the solution in which an internal reference electrode is immersed. In the *nonsymmetrical* arrangement, one side of the membrane is contacted by the sample (usually aqueous) while the other side is contacted by some solid material. An example of this type is provided by coated wire electrodes and ion-sensitive field-effect transistors.

4.1.3.1. Symmetrical Ion-Selective Electrodes

A schematic diagram of a conventional ion-selective electrode was shown in Figure 4-11. The membrane is placed between the sample and the internal reference solution. The composition of the internal solution can be optimized with respect to the membrane and the sample solution. In the interest of symmetry it is advisable to use the same solvent inside both the electrode and sample. This solution also contains the determinand ion in the concentration usually situated in the middle of the dynamic range of the response of the membrane. The ohmic contact with the internal reference electrode is provided by adding the salt which contains the appropriate ion to form a fast reversible couple with the solid conductor. In recent designs, gel-forming polymers have been added to the internal compartment. They do not significantly alter the electrochemistry but add some convenience in handling.

The obvious advantage of a symmetrical arrangement is that the processes at all internal interfaces can be well defined and that most nonidealities at the membrane–solution interface tend to cancel out. The volume of the internal reference compartment is typically a few milliliters, hence the electrode does not suffer from exposure to electrically neutral compounds which would penetrate the membrane and change the composition of this solution. This type of potentiometric ion sensor has been used in the majority of basic studies into ion-selective electrodes. Most commercial ion-selective electrodes are also of this type. The drawbacks of this arrangement are also related to the presence of the internal solution and to its volume. Mainly for this reason it is not conveniently possible to miniaturize and integrate it into a multisensor package.

4.1.3.2. Nonsymmetrical Ion Sensors

The general tendency to miniaturize chemical sensors and make them more convenient for use has led to the development of potentiometric sensors with solid internal contact. These sensors include coated-wire electrodes (CWE), hybrid sensors, and ion-sensitive field-effect transistors. This contact can be a conductor, semiconductor, or even an insulator. The price to be paid for the convenience of these sensors is the more restrictive design rules that must be followed in order to obtain sensors with performance comparable to conventional symmetrical ion-selective electrodes.

The key issue in these sensors is the interface between the ion-selective membrane and the contact. The most convenient way to present this problem is in the form of an equivalent electrical circuit in which the resistances and capacitances have their usual electrochemical meaning (Figure 4-20). Associated with the internal contact is the parasitic capacitance and resistance. It is necessary to include the electrometer (or at least its input stage) in an analysis of these sensors. In most modern instruments it is an insulated-gate field-effect transistor (IGFET) which has the input DC resistance of 10^{14} Ω and the input capacitance on the order of picofarads.

Two limiting cases are treated for the values of the equivalent resistances in Figure 4-20. The charge-transfer resistance R_1 is low for a good ion-selective electrode ($i_0 = 10^{-3}$ A cm^{-2}, i.e., $R_{ct} = 25$ Ω cm^2). This means that at least one charged species can transfer easily between the sample and the membrane and establish the potential difference according to one of the

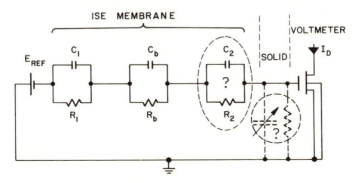

Figure 4-20. Equivalent electrical circuit of a nonsymmetrical membrane arrangement. Subscripts b, 1, and 2 refer to the bulk, membrane/sample, and membrane/solid interface.

mechanisms discussed above. The bulk membrane resistance R_b can be as high as $10^6 \, \Omega \, cm^2$. However, because no net current passes through the membrane the potential in the bulk of the membrane is uniform, i.e., there is no electric field inside the membrane. In a conventional symmetrical ISE arrangement the composition of the internal solution can always be chosen such that the interfacial charge-transfer resistance R_2 is comparable to R_1. Thus, a well-established potential profile exists throughout this structure. On the other hand, if the value of R_2 is infinitely high (i.e., no charge transfer between the internal contact and the membrane), then the input capacitor together with the interfacial capacitance C_2 and the (variable) parasitic capacitance form a (variable) capacitive divider. In such a case the voltage at the input capacitor of the electrometer depends not only on the electrostatic potential of the membrane, but also on the undefined parasitic impedance of the connector. This constitutes a violation of the one-capacitor rule mentioned earlier. In a real situation the interfacial resistance R_2 will have some finite value and the electrode will drift.

a. Coated-Wire Electrodes. A schematic diagram of a coated-wire electrode is shown in Figure 4-21. The internal wire is coated with an ion-selective polymeric membrane, or it can be pressed or glued to the solid-state membrane thus forming a compact and inexpensive ion sensor.[17] A noble-metal internal wire was used in the early designs of CWEs, mainly for its chemical stability. Such electrodes often exhibited unacceptable drift in which the high value of the internal charge-transfer resistance between the membrane and the wire was soon implicated. This is not surprising if it is realized that the typical electrode reactions at noble metals are of redox type, i.e., the charge-transferring species is the electron. On the other hand, in most ion-selective membranes the charge conduction is carried out by ions. Thus, a mismatch between the charge-transfer carriers can exist at the noble-metal/membrane interface. This is especially true for polymer-based membranes, which are invariably ionic conductors. However, solid-state membranes which exhibit mixed ionic and electronic conductivity (such as chalcogenide glasses, petrovskites, and silver halides) form a good contact with noble metals.

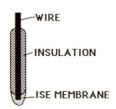

Figure 4-21. Diagram of a coated-wire electrode.

There have been several approaches to solving the problem of the internal contact. The most direct approach involves the interposition of a thin layer of aqueous gel which contains the fixed concentration of the salt of the primary ion. This approach has met with only limited success and can be seen as the miniaturization of the conventional ISE structure. The main reason for its failure is that electrically neutral species can permeate through the membrane and establish their own activities inside the sensor structure according to their chemical potential in each individual phase. Therefore, the water permeating through the membrane reaches its osmotic equilibrium in line with the concentration of the solutes present in the gel. This can lead not only to a significant change in the internal activity of the primary salt inside the gel, but often to the catastrophic failure of the whole structure when the osmotic pressure inside the gel exceeds the limits of the mechanical stability of the sensor. This problem is difficult to avoid, because the concentration of otherwise neutral solutes in the sample is not known *a priori* and cannot be controlled during the measurement.

The second problem relates to Severinghaus-type interference (see Section 4.1.5.1 below). When an electrically neutral chemical species with acid/base or redox properties penetrates the membrane (e.g., acetic acid or iodine), it can dissociate inside the gel, change its pH and/or redox potential, and thus affect the interfacial potential.[18] In principle, this problem can be avoided by establishing an electrochemical process with as high an exchange-current density as possible at this interface. It is hoped that this process will dominate the interfacial potential, which will become independent of any small changes that may occur due to the above interference. In principle the same effect can cause problems in conventional ISE, however, it does not happen in practice mainly due to the relatively large volume (e.g., 3–5 ml) of the internal solution.

It has been found that a much more stable CWE can be made by replacing the noble metal with copper or silver. The use of a thin coating of AgCl on top of the silver wire is particularly popular. An internal contact made of carbon should suffer from the same problems as the noble-metal CWE. Such a contact exists in Selectrodes[19] in which the components of an ion-selective membrane are mixed with carbon paste. Carbon has similar electrochemical properties to those of noble metals. However, contact with the membrane is made over a very large area far exceeding the geometrical surface area of the electrode. This interface is substantially stabilized because the interfacial resistance is inversely proportional to the contact area.

The fluoride ion-selective electrode used in the cell discussed in Section 4.1.3 is another example of a nonsymmetrical potentiometric sensor. It cannot properly be called a coated-wire electrode because contact between the membrane and internal connector has been made not by coating but by some thin-film deposition technique (such as sputtering). Nevertheless, the electrochemistry involved in the operation of this sensor is quite the same. It is again essential to identify the charge-transferring species at each interface. In this case AgF is used instead of the usual AgCl, because fluoride ion is needed to couple the AgF and LaF$_3$ layers. The transformation from ionic to electronic transport takes place at the Ag/AgF interface. The solid contact between a regular pH glass electrode and silver wire has also been made by using AgF as intermediate ionic conductor.[20] In that case the coupling species seems to be the OH$^-$ ion.

In summary, it is perfectly legitimate to design CWEs and expect an electrochemical performance comparable to the conventional ISE. However, the membrane–solid interface must be designed with an understanding of the electrochemical processes at such an interface. If the final resistance is high, then problems such as drift become directly proportional to the length of the internal contact, specifically to the magnitude of the parasitic capacitance and resistance. It can therefore be concluded that these problems can be minimized by decreasing the distance between the ion-selective membrane and the amplifier. The logical stage in the development of these sensors is hybrid ion sensor[20] and an ion-sensitive field-effect transistor (Figure 4-22).

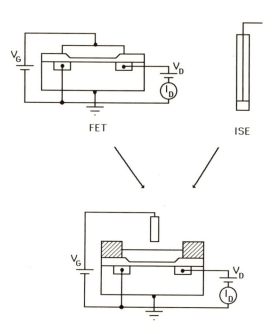

Figure 4-22. Transition from IGFET and ISE to ISFET.

b. Field-Effect Transistors. Field-effect transistors are part of any modern pH meter. With the introduction of ion-sensitive field-effect transistors they have come to the attention of chemists. In order to understand the principles underlying the operation of these new electrochemical devices it is necessary to include FET in the overall analysis of the measuring circuit. The rules governing the operation of these semiconductor devices will first be outlined briefly.

The semiconductor field effect.[21] In an extrinsic semiconductor containing only one type of dopant there are equal densities of mobile charges (electrons in n-type and holes in p-type semiconductors) and ionized dopant atoms (positively charged for n-type and negatively charged for p-type semiconductors). For simplicity this discussion is restricted to a p-type extrinsic semiconductor. The condition of electroneutrality applies only in the absence of an external electric field.

If an electric field is applied to the surface of the semiconductor from whatever source, the density of the mobile charge carriers (holes) will be either enhanced or depleted depending on the polarity of the field. If the field enhances the concentration of holes, the surface is said to be accumulated and the semiconductor surface behaves much like a metal in that the excess charge appears at the surface and the electric field does not penetrate it further. If, on the other hand, the field forces the mobile holes away from the surface, a space-charge region consisting of the ionized acceptor atoms, which are fixed in the lattice, forms over an appreciable distance into the semiconductor. The thickness of the surface space-charge region is a function of the strength of the field at the surface and the semiconductor doping profile, as is the difference between the surface and bulk potentials of the semiconductor. If the surface potential deviates sufficiently far from the bulk potential, the surface will invert, that is, it will contain an excess of mobile electrons. In this case there is an n-type conductive channel on the surface separated from the p-type bulk by a space-charge region. Further increase in the normal surface field will not significantly change the surface potential, but will only change the density of electrons in the n-type surface layer and, consequently, the electrical conductivity of this layer. Changes in the surface electric field can be determined by measuring the corresponding changes in the thickness of the surface depletion region if the surface is not inverted, or the conductivity of the surface inversion layer if the surface is inverted.

The surface field can be produced in a number of ways. The semiconductor can be built into a capacitor and an external potential applied (IGFET), or the field can arise from the electrochemical effects between different materials (ISFET). In both cases, variations in the surface electric field change the density of mobile charge carriers in the surface inversion layer. The physical effect measured is the change in the electric current carried by the surface inversion layer, the drain current. The devices must be operated under conditions that cause the surface inversion layer to form.

Current–voltage relationships for the IGFET. Figure 4-23 shows the typical construction and biasing arrangement for the n-channel IGFET. In effect the semiconductor substrate (3) forms one side of a parallel-plate capacitor and the metal-gate field plate (4) forms the other. In the normal operating mode the gate voltage V_G is applied between the semiconductor substrate (3) and the gate (4). The polarity and magnitude of V_G are such that the semiconductor field effect gives rise to an inversion layer on the surface of the substrate under the gate. In fact, the surface of the p-type substrate (3) becomes n-type. This n-type inversion layer forms a conducting channel between the source (2) and drain (1) regions. If this inversion layer is not present, application to the drain (1) of a positive voltage V_D with respect to the source (2) and substrate (3) results in no appreciable current flow into drain (1) because the drain-to-substrate p–n junction diode is reversed biased. However, if the n-type inversion layer exists along the surface of the semiconductor between the source (2) and drain (1), a continuous path for the flow of current I_D exists. The current I_D flows through the channel connecting the drain (1) and source (2). The magnitude of the drain current I_D will be determined by the electrical resistance of the surface inversion layer and the voltage difference V_D between the source and drain.

The basis for the derivation of the current–voltage relationships is the calculation of the density of the mobile electrons in the surface inversion layer as a function of the applied voltages V_G and V_D, and the position along the channel.

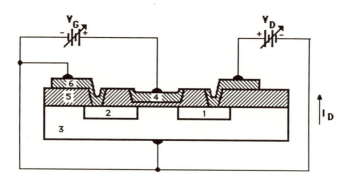

Figure 4-23. Diagram of the n-channel IGFET: (1) drain, (2) source, (3) p-type substrate, (4) metal gate, (5) insulator, (6) metal contacts.

Figure 4-24 is a schematic representation of the channel region of the IGFET. At a point y along the channel there is a charge density $Q_n(y)$ per unit area of mobile electrons in the surface inversion layer, and a charge density $Q_B(y)$ of ionized immobile dopant atoms in the space-charge region. These charge densities are both a function of y as a result of the voltage change along the channel due to the drain current. Quantity Q_s, the sum of $Q_n(y)$ and $Q_B(y)$, is equal and opposite to the charge per unit area on the metal plate which forms the other side of the capacitor (the gate),

$$Q_s = Q_n(y) + Q_B(y) \qquad (4\text{-}57)$$

The value of Q_s may be related to the applied gate voltage and the capacitance C_0 of the gate insulator by

$$V_G - V_{FB} = -Q_s/C_0 + \phi_s \qquad (4\text{-}58)$$

where V_{FB} is the "flat band" voltage defined by equation (4-93). It accounts for the effects of the work function difference between the gate metal and the silicon and any charges that exist in the gate insulator; ϕ_s is the surface potential of the semiconductor. The mobile electron charge in the surface inversion layer is then seen to be given by

$$Q_n(y) = -[V_G - V_{FB} = -Q_s/C_0 + \phi_s(y)]\,C_0 - Q_B(y) \qquad (4\text{-}59)$$

In writing these equations it is assumed that a surface inversion layer does in fact exist. At this point in the derivations the assumption is made that the surface potential is given by the conditions of strong inversion,

$$\phi_s(y) = V(y) + 2\phi_F \qquad (4\text{-}60)$$

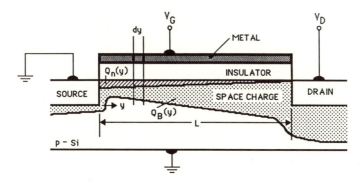

Figure 4-24. Voltage and charge distribution in the channel region of an IGFET.

where $V(y)$ is the "reverse bias" applied in the perpendicular direction to the "field-effect induced" p–n junction composed of the n-type surface inversions layer and the p-type substrate, while $Q_B(y)$ is the charge in the space-charge region. Under the "depletion approximation," it is assumed that there are no mobile charges at all in the space-charge regions,

$$Q_B(y) = -\{2\varepsilon_{Si}\varepsilon_0 q N_A[V(y) + 2\phi_F]\}^{1/2} \tag{4-61}$$

It is seen that $Q_B(y)$ increases with $V(y)$, the voltage drop along the channel. This increase is at the expense of $Q_n(y)$, the mobile channel charge. For the sake of simplicity it is possible to neglect the dependence of $Q_B(y)$ on the voltage drop [see equation (4-61)] which only marginally affects the accuracy of these equations. In that case

$$Q_n(y) = -[V_G - V_{FB} = -Q_s/C_0 + \phi_s(y)] \, C_0 + 2(\varepsilon_{Si}\varepsilon_0 q N_A \phi_F)^{1/2} \tag{4-62}$$

Now referring to Figure 4-24, the voltage drop along the element of channel, dy, can be expressed in the form

$$dV = I_D \, dR = I_D \frac{dy}{W\mu_n Q_n(y)} \tag{4-63}$$

where W is the width of the channel and μ_n is the electron mobility in the channel.

Substitution of the above expression for $Q_n(y)$ and integration over the channel length yields the basic current–voltage relationship

$$I_D = \frac{\mu_n C_0 W V_D}{L} (V_G - V_T - V_D/2) \tag{4-64}$$

where

$$V_T = V_{FB} + 2\phi_F - Q_B/C_0 \tag{4-65}$$

In the derivation of equation (4-64) it was assumed that the surface inversion layer actually existed at all points along the channel. The value of $V(y)$ must change from 0 at the source to V_D, so the conducting channel will disappear near the drain end of the channel if V_D is large enough. When this happens equation (4-64) is no longer valid and the IGFET is said to become "saturated." Figure 4-25 is a schematic diagram showing the saturated IGFET. Quantity V_{Dsat} is the drain voltage at which the surface channel just disappears at the drain end,

$$V_{Dsat} = V_G - V_T \tag{4-66}$$

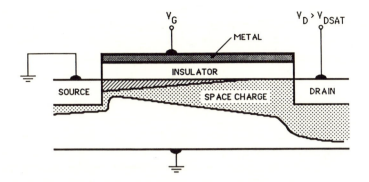

Figure 4-25. Voltage and charge distribution in the IGFET operated in saturation.

Any increase in V_D beyond V_{Dsat} results in a short space-charge region between the channel and the drain. The drain current continues to flow under these conditions, because electrons in the channel see no potential barrier restricting flow from the channel across the space-charge region to the drain. However, the number of electrons arriving at the drain end of the channel is determined by the voltage V_{Dsat} between the source and the end of the surface channel. For drain voltages greater than V_{Dsat}, the drain current I_D is given by equation (4-67) with V_D replaced by V_{Dsat}, provided the channel length L is not appreciably shortened by the space-charge region between the end of the channel and the drain,

$$I_D = \frac{\mu_n C_0 W}{2L} (V_G - V_T)^2 \qquad (4\text{-}67)$$

The curves in Figure 4-26 are divided into saturated and unsaturated regions.

Equations (4-64) and (4-67) have the distinct advantage of possessing a simple form and are the IGFET current–voltage relations most often quoted in the literature. Their use can lead to significant quantitative errors in calculated currents. They do, however, contain the correct qualitative features of the device, i.e., the saturated and unsaturated regions, the near-constant drain current beyond saturation, and the dependence of the saturation voltage on the gate and threshold voltages.

Figure 4-27 is a plot of I_D vs V_G for an n-channel depletion-mode device. Depletion mode refers to the fact that the device has a conducting channel for zero applied gate voltage. It is the result of a sufficiently negative flat-band voltage, V_{FB}. In the saturated regions, i.e., $V_D > V_G - V_T$, or alternately $V_G < V_D + V_T$, equation (4-67) predicts a quadratic dependence of I_D on V_G as shown. In the unsaturated region, i.e.,

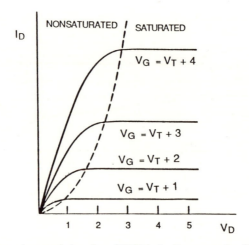

Figure 4-26. Output characteristics of an IGFET: I_D–V_D curves for different values of the gate voltage.

$V_D < V_G - V_T$, or alternately $V_G > V_D + V_T$, equation (4-65) predicts a linear dependence of I_D on V_G as shown. Actual data, however, show significant departure from linearity in the unsaturated region as indicated in Figure 4-27. One cause of this departure from theory is the variation in μ_n, the charge-carrier channel mobility, with the electric field normal to the surface of the semiconductor. Surface channel mobility is primarily determined by surface scattering. Strong electric fields that increase the

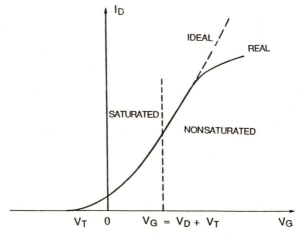

Figure 4-27. Drain current (I_D)–gate voltage (V_G) curves for ideal and real depletion mode IGFET. Departure from ideality is caused by the approximate nature of equations (4-64) and (4-67), and by the resistance in series with the channel.

probability of the carrier interacting with the surface reduce the surface mobility. Therefore, as the gate voltage increases, the electric field normal to the surface increases and μ_n decreases. A second cause of the departure of the I_D–V_G curve from linearity is series resistance between the end of the channel and the point at which the drain voltage is actually measured. The resistive voltage drop actually reduces the effective drain voltage below that applied to the device. The magnitude of this resistance depends on both the individual device geometry and processing parameters, but it may be quite significant, up to 200 Ω in some designs.

As a result of the effects of nonideal structures, second-order effects in parameters, and the numerous approximations made in the derivation of the current–voltage equations, equations (4-64) and (4-67) can only serve as a qualitative description of the actual device; each individual design must be experimentally characterized. For these reasons it is advantageous to operate the FET in the constant drain-current mode, in which case a suitable feedback circuit supplies the gate voltage of the same magnitude but of opposite polarity to that produced by the electrochemical part of the device.

The fact that IGFET can be perceived as a field-dependent resistor is seen from the expression for transconductance g_m, which is defined as the average conductance of the channel for a given value of V_D. For non-saturation, from equation (4-64)

$$g_m = \left| \frac{I_D}{V_G} \right|_{V_D} = \mu_n C_0 W V_D / L \qquad (4\text{-}68)$$

and for saturation, from equation (4-67)

$$g_m = \left| \frac{I_D}{V_G} \right|_{V_D} = \mu_n C_0 W (V_G - V_T) / L \qquad (4\text{-}69)$$

It is useful to summarize the assumptions made in the derivation of equations (4-64) and (4-67):

1. It is assumed that there exists a well-defined threshold voltage, and that the formation of the surface inversion layer begins suddenly as the gate voltage is increased. It is equivalent to stating that there is a sharply defined semiconductor surface potential dividing surface depletion and surface inversions. In fact, this transition is continuous. In conventional structures this is a good approximation if the gate voltage exceeds the "threshold voltage" by about 0.56 V.

2. The voltage drop due to a surface channel current (drain current) flow has no effect on the thickness of the surface space-charge region. This approximation can lead to relatively large errors in the magnitude of the predicted drain current in the saturation region, but the general shape of the drain current vs. drain voltage curves is satisfactory, i.e., the qualitative features of the device are unaffected by this approximation.

3. Doping of the semiconductor is constant near the surface where the channel is formed. This is not a good approximation for many ion-implanted structures.

4. The channel length is large compared to the thickness of the depletion region surrounding the p–n junctions.

5. Both the source and substrate of the device are connected to the same point of the external circuit.

6. The thickness of the space-charge region along the channel is constant.

c. Ion-Sensitive Field-Effect Transistors. The idea of the first truly integrated chemical sensor, the ion-sensitive field-effect transistor (ISFET),

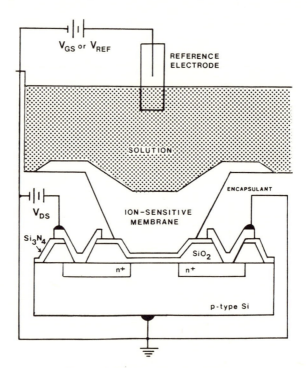

Figure 4-28. Diagram of an ISFET.

was introduced independently by Bergveld[22] and by Matsuo[23] in the early 1970s. The metal gate of an ordinary IGFET was removed and the gate insulator directly exposed to the electrolyte solution. It will be shown later that there is a pH-dependent charge at the insulator/solution interface which contributes to the surface charge [see equation (4-57)].

A schematic diagram of an ISFET is shown in Figure 4-28. The metal gate is replaced by a chemically sensitive layer. The electrical path is closed with a reference electrode and the conducting solution, while the device is protected by a suitable encapsulant. In this section two types of chemically sensitive field-effect transistor (CHEMFET) will be considered: one utilizing an ion-selective membrane (ISFET) and one utilizing an enzyme layer to achieve chemical selectivity (ENFET).

The heart of an ISFET is the gate. It has been shown in the preceding section how the gate voltage V_G controls the drain current I_D in the transistor. It was thought initially that ISFETs can be operated without a reference electrode. Such an operation would violate the established electrochemical laws, but it seemingly makes sense from the viewpoint of the operation of the transistor. In order to understand this argument it is necessary to perform the thermodynamical analysis of the structure shown in Figure 4-29, which represents the cross-section through the whole measuring circuit including the reference electrode and connecting leads. It

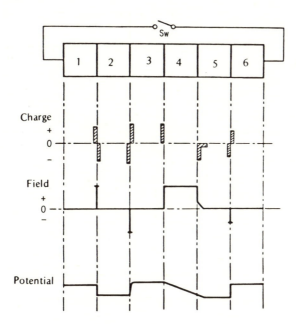

Figure 4-29. Cross-section, charge, field, and potential distribution in the ISFET gate.

is the simplest case, one where the reference electrode (1) is of the first kind, described by the equilibrium

$$M = M^+ + e^-$$

The solution (2) is assumed to also contain a small amount of ions that can permeate reversibly into and out of the membrane (3), which therefore forms a nonpolarized interface. A possible example would be a solution (2) containing 0.1 M $AgNO_3$ and 0.1 mM KNO_3 with the membrane (3) being potassium-ion-sensitive, such as used in the equivalent ISE. The insulator (4) is assumed to be ideal, i.e., no charge can cross it and it is thicker than the electron tunneling distance ($d > 100$ Å). Layer (5) is the transistor semiconductor substrate. The metal (6) is identical with metal (1). A switch S_w represents operation with (S_w closed), and without (S_w open) a reference electrode. The charge, field, and potential profiles across this structure are also shown in Figure 4-29. It should be noted that this is a very simplified case. A liquid junction, dual-layer insulator, trapped charges in the insulator, surface states at the insulator–semiconductor interface, channel doping profile, and a multitude of connecting metals have been omitted for the sake of simplicity.

From the thermodynamic point of view, this is a multiphase system for which the Gibbs equation (A-20) must apply with equilibrium at each interface. There is no charge transfer in and out of layer (4) (ideal insulator), so the sandwich of layers (3)/(4)/(5) represents an ideal capacitor.

It follows from the Gibbs equation that this system can reach electrostatic equilibrium by closing the switch S_w. If the switch S_w is open, another capacitor (1)/(?)/(6) is formed. The "questionmark ?" signifies the undefined nature of this capacitor. This situation is equivalent to operation without a reference electrode (or a signal return). An acceptable electrostatic condition would be reached only if the second capacitor had a defined and invariable geometry. This is again the "one-capacitor" rule discussed previously in Section 4.1.

The circuit in Figure 4-29 will now be analyzed. The inner potential in semiconductor (5) can be expressed as

$$\phi_5 = (\mu_5^e - \tilde{\mu}_5^e)/F \tag{4-70}$$

where μ_s^e is the chemical potential of the electron in the semiconductor (the electron–lattice interaction energy) and $\tilde{\mu}_5^e$ is the electrochemical potential of the electron in phase 5, normally known as the Fermi level. Similarly, for membrane (3) the inner potential ϕ_3 is given by

$$\phi_3 = (\tilde{\mu}_3^i - \mu_3^i)/z^i F \tag{4-71}$$

where z_i, the number of elementary charges, is positive for cations and negative for anions, while $\tilde{\mu}_3^i$ and μ_3^i are the electrochemical and chemical potentials of species i in phase 3, respectively. The potential difference across the insulator and semiconductor space-charge region is then

$$\phi_5 - \phi_3 = (\mu_5^e - \tilde{\mu}_5^e)/F - (\mu_3^i - \tilde{\mu}_3^i)/z^i F \qquad (4\text{-}72)$$

It is now essential to identify the relationship between species i in membrane (3) and the electron in semiconductor (5). We know that ion i can transfer from solution (2) into membrane (3), thus according to equation (4-72) its electrochemical potentials in the two phases must be equal:

$$\tilde{\mu}_3^i = \tilde{\mu}_2^i = \mu_2^i + z^i F \phi_2 \qquad (4\text{-}73)$$

where ϕ_2 is the inner potential in the solution. Similarly, Fermi levels in the semiconductor and metal (6) are equal. Metal (6) was defined to be the same as metal (1) (the reference electrode), hence we can write

$$\tilde{\mu}_5^e = \tilde{\mu}_1^e = \mu_1^e - F \phi_2 \qquad (4\text{-}74)$$

There is equilibrium in the metal between the metal cations and electrons, namely

$$M_1 = M_1^+ + e_1^-$$

for which we can formally write

$$\mu_1^M = \mu_1^{M+} + \mu_1^e \qquad (4\text{-}75)$$

On substituting for μ_1^e in equation (4-74) we obtain

$$\tilde{\mu}_5^e = \mu_1^M = \mu_1^M - F \phi_1 \qquad (4\text{-}76)$$

If equations (4-72), (4-74), and (4-76) are combined and rearranged, then

$$\phi_5 - \phi_3 = (\mu_5^e - \mu_1^M + \mu_1^{M+})/F - (\mu_2^i - \mu_3^i)/z^i F + \phi_1 - \phi_2 \qquad (4\text{-}77)$$

The first term on the right-hand side of equation (4-77) combined with equations (4-74) and (4-75) yields the contact potential between the semiconductor and metal,

$$\phi_5 - \phi_1 = (\mu_5^e - \mu_1^e)/F = \Delta\phi_{\text{cont}} \qquad (4\text{-}78)$$

The second term is related to the solution activity of the ion by the Nernst equation

$$(\mu_2^i - \mu_3^i)/z^i F = E_0^i + (RT/z^i F) \ln a_2^i \qquad (4\text{-}79)$$

where a_3^i is assumed constant and incorporated in the term E_0^i. Finally,

$\phi_1 - \phi_2$ is the reference electrode potential E_{ref}. Equation (4-74) now assumes the form

$$\Delta\phi_{3/5} = \Delta\phi_{\text{cont}} + E_0^i + (RT/z^iF) \ln a_2^i - E_{\text{ref}} \qquad (4\text{-}80)$$

The voltage $\Delta\phi_{3/5}$ across the gate insulator is added to the externally applied voltage V_G and the resulting voltage has the same meaning and function as defined in the theory of IGFET operation. For ISFET operated in nonsaturation conditions, equations (4-64) and (4-80) yield

$$I_D = \frac{\mu_n C_0 W V_D}{L} (V_G - V_T + \Delta\phi_{\text{cont}} + E_0^i + (RT/z^iF) \ln a_2^i - E_{\text{ref}} - V_D/2)$$

$$(4\text{-}81)$$

We define V_T^* for ISFET as

$$V_T^* = V_T - \Delta\phi_{\text{cont}} - E_0^i \qquad (4\text{-}82)$$

The inclusion of the term E_0^i (but not E_{ref}) in the newly defined threshold voltage is rather arbitrary. The reason here is that the membrane is physically part of the ISFET and thus its standard potential should be included in the constant V_T. On the other hand, the reference electrode is a completely separate structure and therefore its potential is kept separate. The final equation for the drain current of ISFET sensitive to the activity of ions i in the nonsaturation region is then

$$I_D = \frac{\mu_n C_0 W V_D}{L} [V_G - V_T^* \pm (RT/z^iF) \ln a_2^i - E_{\text{ref}} - V_D/2] \quad (4\text{-}83)$$

For operation in the saturation mode we obtain, from equations (4-67), (4-80), and (4-82),

$$I_D = \frac{\mu_n C_0 W}{2L} [V_G - V_T^* \pm (RT/z^iF) \ln a_2^i - E_{\text{ref}}]^2 \qquad (4\text{-}84)$$

Equations (4-83) and (4-84) relate the activity of the measured ion in the sample to the output of ISFET in a rather inconvenient way. The fact that these equations have been derived with the help of some drastic assumptions makes them only semi-quantitative. Furthermore, whether the actual output follows equation (4-83) or (4-84) depends on the applied gate voltage V_G, which has nothing to do with the electrochemistry involved. Mainly for these reasons it is preferable to operate the ISFET in the so-called "feedback mode." By using an operational amplifier in a feedback

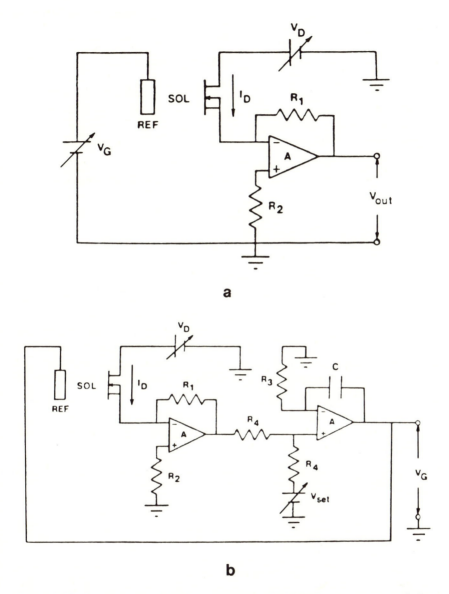

Figure 4-30. Two circuits used for the operation of a CHEMFET: (a) constant applied gate voltage corresponding to equation (4-83); (b) the more explicit operation at constant drain current (a feedback operation) which yields directly the Nernst potential. (Reprinted from Ref. 21.)

loop as shown in Figure 4-30, the operating point of the ISFET may be held constant. Any change in the potential difference between the solution and the ion-selective membrane due to a change in ionic activity i appears as a change in the applied potential V_G. Naturally, it is possible to realize the feedback circuit in many different ways, in either analog or digital form. The basic principle is the same. In the circuit of Figure 4-30, the operational amplifier will maintain the voltage V_{out} at whatever value is required to make I_D equal to I_{Dset}. In response to a change in the activity of the measured ions, it will automatically adjust V_G in equation (4-83) or (4-84) to balance the change due to the term $(RT/z^iF)\ln a_2^i$. Since nothing else varies, the readout is directly in mV as is customary in conventional potentiometric measurements.

The *pH ISFET* will now be examined. Close integration of the selective layer with the amplifer, as it occurs in CHEMFET, offers some unique sensing possibilities. We must remember that the field-effect transistor is basically a charge-sensing device. From the very beginning of research on this device it has been known that the FET with bare-gate insulator exposed to the solution responds to changes in pH. This quantity is such an important parameter to be measured that a very considerable amount of research has been devoted to explaining the mechanism of this response. Clearly, there is a pH-dependent charge at the solution–transistor interface. The slope of this response has been found to vary between 50% of the theoretical value for SiO_2 to 92% for silicon nitride. Other materials, such as Al_2O_3 or Ta_2O_5 and other oxides, have been reported to yield a nearly theoretical response. It is necessary to pause here and realize that all these materials are very good bulk insulators. Therefore, they could not be used as pH-sensing membranes in a conventional ISE configuration because they would create another capacitor in the circuit and violate the "one-capacitor" rule. However, in an ISFET configuration these materials *are* actually part of the *one capacitor*, the input gate capacitor, and their use is possible.

It is quite adequate to accept the fact that the pH-dependent charge resides somewhere at the insulator "surface" and that the corresponding image charge in the transistor channel affects the transconductance. We now invoke the basic electrochemical characteristics of the interface, the distinction between the nonpolarized (resistive) and polarized (capacitive) interface (Section 4.1). The argument about the location of the pH-dependent charge has revolved around this distinction: one school of thought has been that the charge is located in one plane and that the interface behaves as a capacitor at which the charge is generated by the de-protonation/protonation of the surface bound sites. This is the basis of the "site-binding theory" (SBT). In this theory it is assumed that there are ionizable binding sites present at the surface of the insulator and that they determine the distribution of the compensating charge in the adjacent layer of solution. This distribution is the result of a combination of the interplay between the thermal forces and electrostatic forces originating in the fixed pH-dependent charge at the surface as governed by the Poisson–Boltzmann distribution. The potential decays exponentially from its surface value ϕ_0 according to

$$\phi_x = \phi_0 \exp(-\kappa_D x) \qquad (4\text{-}85)$$

with the effective thickness κ_D^{-1} of the space-charge region defined as

$$\kappa_D = (8\pi C_0 z^2 e_0^2/\varepsilon kT)^{1/2} \qquad (4\text{-}86)$$

where ε is the dielectric constant, C_0 the bulk concentration, and e_0 the charge of the electron.

The other model postulates the existence of the hydrated layer of finite thickness within which the de-protonated/protonated sites were situated. The creation of the boundary potential follows an argument similar to that presented in Section 4.1.1.1. Clearly, in the limit of zero thickness of the hydrated layer the two theories merge. It is obvious that the SBT models the interface as a capacitor with one plate located at the surface and the other at the average distance κ_D^{-1} in the solution. In other words, the interface is considered to be ideally polarized. On the other hand, the hydration layer model allows penetration of at least one type of ion through the interface and thus regards this interface as *nonpolarized*. The degree of nonpolarization depends on the value of the exchange-current density of the communicating ion(s). The equivalent electrical model is, of course, a parallel capacitor/resistor combination (Figure D-1 in the Appendix). From the standpoint of the response it is obvious that the interface described by the capacitve model (SBT) would respond to adsorption of any charged species inside the space-charge region. On the other hand, adsorption would have little or no effect on the potential difference at the interface described by the hydration layer model, because that potential difference is uniquely and unequivocally given by the Gibbs equation. Of course, a complicating factor in a real situation would be the existence of the mixed potential (Section 4.1.1.4) at the interfaces, which have low total exchange-current density.

The common ground to accommodate both models has been identified by Sandifer,[24] who showed that the sub-Nernstian response of some materials and the presence of troublesome adsorption and stirring effects can be rationalized by considering the number of available binding sites and the thickness of the hydrated layer. The magnitude of the exchange-current density depends on the concentration of the binding sites inside the

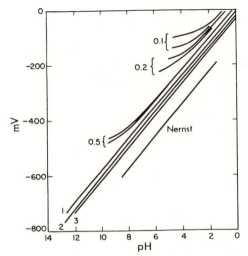

Figure 4-31. Dependence of the Nernst potential on the density of the binding sites and on adsorption. Upper curve in the bracketed sets corresponds to the absence of adsorption. The lower curves show the effect of 100 mM charged adsorbate. (Reprinted from Ref. 24 with permission of the American Chemical Society.)

hydrated layer. For a glass electrode this is estimated to be 3.2 M.[13] As the number of binding sites decreases from 3 to 0.1 M, the interfacial potential of a 10-Å-thick layer is increasingly affected by adsorption (Figure 4-31). At the same time the response deviates from theoretical the less the binding sites are present. However, both the adsorption effects and

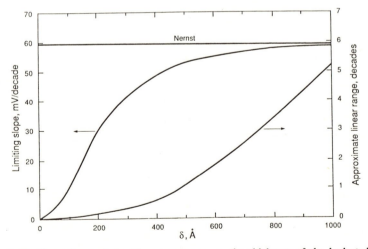

Figure 4-32. Dependence of the Nernst response on the thickness of the hydrated layer. (Reprinted from Ref. 24 with permission of the American Chemical Society.)

the sub-Nernstian behavior vanish if the thickness of the hydrated layer is allowed to increase up to 800 Å (Figure 4-32).

This model goes a long way toward explaining some experimental results reported in the literature for ISFETs with oxide or nitride surfaces. Unfortunately, the properties of these materials, prepared in different laboratories, differ substantially. It is known that silicon nitride forms an oxygen-rich layer at the surface whose thickness depends on the deposition conditions. This "passivation" layer seems to form rapidly, but is very stable even under continuous exposure to aqueous electrolyte solutions. The hydration of this layer seems to fit the requirements and predicted behavior of the Sandifer model. As expected, ISFETs which have been exposed to solution for a minimum of one hour show no adsorption effects that would be expected from the SBT model.[21]

4.1.4. Enzyme Electrodes and Enzyme Transistors

From the point of view of the diffusion-reaction kinetics, these sensors belong to the zero-flux boundary group together with enzymatic thermal and enzymatic optical sensors. In fact the proposed model, which was discussed in Section 1.2.2, has been verified on enzymatic field-effect transistors for the diffusion-reaction mechanism involving the oxidation of β-D-glucose catalyzed by a glucose oxidase (GOD)/catalase system[25] (Scheme 4-1) and for hydrolysis of penicillin catalyzed by β-lactamase[26] (Scheme 4-2).

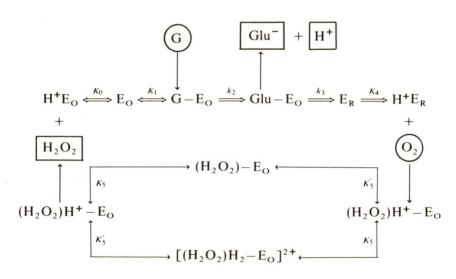

Scheme 4-1. Glucose oxidation.

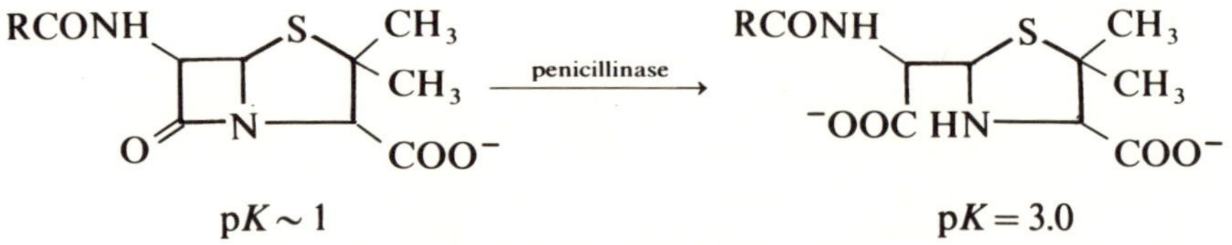

Scheme 4-2. Penicillin hydrolysis.

In both cases hydrogen ion is the species detected by the transistor surface. The experimental and calculated response according to the above model is shown in Figures 4-33 through 4-36. It was pointed out earlier that their agreement, which verifies the validity of the model, allows us to plot the concentration profiles inside the gel layer (Figures 4-37–4-39). It is quite evident that the concentration gradients are not linear either in the steady state or during the transition period. This shows that the assumption of the linear concentration gradient used in the early models of potentiometric enzymatic sensors is not justified. Other important information provided by the concentration profiles is the value of the minimum required thickness. As expected from equation (1-42), it depends on the value of the Thiele modulus which in turn depends on the reaction mechanism. The required thickness is approximately 150 μm for the glucose sensor (Figure 4-37) and 20 μm for the penicillin one (Figure 4-39). From these profiles we can see that the thinner membranes or directly immobilized enzymes would produce a lower sensitivity. On the other hand, the thicker the enzyme layer the slower the response. Thus, we can trade the time response for the signal-to-noise ratio by adjusting the thickness of the enzyme-containing layer.

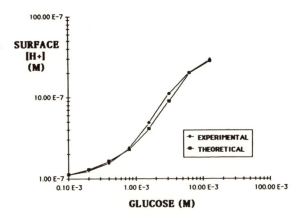

Figure 4-33. Calibration curve for glucose-sensitive ENFET in 0.2 mM phosphate buffer (pH 7.2) saturated with oxygen at 300 °C. (After Ref. 25.)

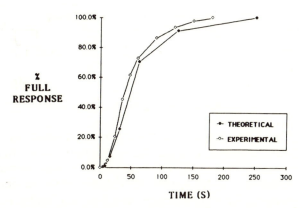

Figure 4-34. Time response of a glucose ENFET for a concentration step from 0 to 1 mM glucose under conditions given in Figure 4-33.

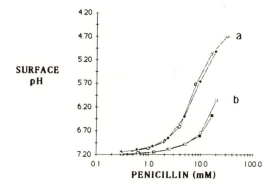

Figure 4-35. Calibration curve for penicillin-sensitive ENFET. Two concentrations of pH 7.2 phosphate buffer were used: (a) 20 mM and (b) 80 mM. (After Ref. 26.)

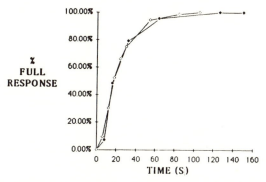

Figure 4-36. Time response of a penicillin ENFET for a concentration step from 0 to 10 mM penicillin in 20 mM pH 7.2 phosphate buffer. (After Ref. 26.)

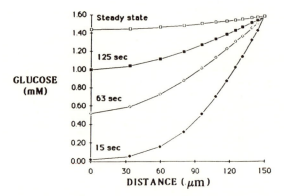

Figure 4-37. Concentration profile for glucose through a 150-μm-thick membrane under conditions given in Figure 4-33. The initial concentration of glucose was 1.5 mM. (After Ref. 25.)

Other parameters affecting the response characteristics of these sensors are the partitioning coefficients and the diffusion coefficients of all species, and the parameters related to the enzyme activity itself, such as the enzyme concentration, K_m, and v_m. They are in turn affected by the preparation of the enzyme layer, e.g., by the degree of crosslinking, by the degree of the inactivation of the enzyme. It is therefore not surprising to find that these devices have generally widely different lifetime, time response, detection limits, and sensitivity. The value of the experimentally verified model is then mainly in establishing the design parameters not only for potentiometric but also for other zero-flux boundary enzymatic sensors.

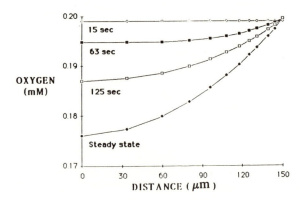

Figure 4-38. Concentration profile for oxygen through a 150-μm-thick membrane. Same conditions as in Figure 4-37.

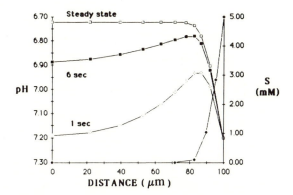

Figure 4-39. Concentration profile for penicillin through a 100-μm-thick membrane for a concentration step from 0 to 5 mM penicillin in 2 mM pH 7.2 phosphate buffer. (S. Caras, unpublished results.)

Enzymes provide such an attractive possibility for achieving chemical selectivity, so enzyme electrodes were the first enzymatic chemical sensors (or first biosensors) to be made. The early designs used any available method for immobilization of the enzyme at the surface of the electrode. Thus, physical entrapment using dialysis membranes, meshes (Figure 4-40), and various covalent immobilization schemes have been employed (Section 1.2.4). The enzyme-containing layer is simply added onto the existing ion sensor. The use of ISFETs enabled precise control over the geometry of the enzyme layers as well as the miniaturization of these devices (Figure 4-41).

The choice of ion sensor clearly depends on the type of enzymatic reaction, namely, on the products and reagents of that reaction and on the conditions of the sample. Hence, for example, there are several possibilities of choosing the ion sensor for an enzyme electrode for urea.

$$NH_2 - CO - NH_2 + H_2O \xrightarrow{\text{urease}} 2NH_3 + CO_2$$

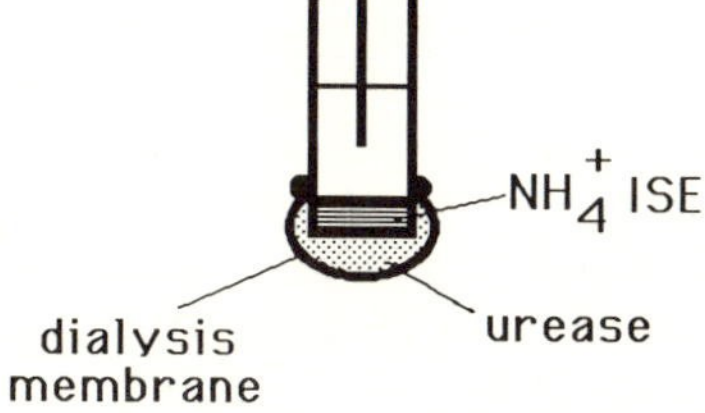

Figure 4-40. Schematic diagram of a potentiometric urea electrode.

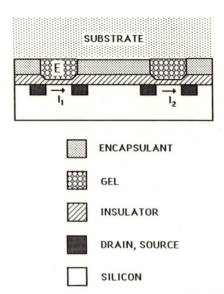

Figure 4-41. Diagram of a dual-gate enzymatic field-effect transistor.

One could immobilize the urease layer on top of a Severinghaus electrode for CO_2 or NH_3 (Section 4.1.5.1) and use the device as an enzymatic–potentiometric gas sensor. The primary disadvantage of such an arrangement would be a slow time response. A more direct way is through detection of the ionic species resulting from the hydrolysis of ammonia and carbon dioxide,

$$NH_3 + H_2O \Leftrightarrow NH_4^+ + OH^-$$

and

$$CO_2 + H_2O \Leftrightarrow HCO_3^- + H^+$$

An ionic sensor for any of the ions in the above equilibria can be used in this case. The choice is dictated usually by the conditions of the sample. Thus, at low pH values the species of choice would be NH_4^+ and at high pH values it would be HCO_3^-. For a neutral pH range the best performance would be obtained from monitoring the changes of pH itself.

Since each enzyme sensor has its own unique response, it is necessary to construct the calibration curve for each sensor separately. The obvious way to reduce interferences is to use two sensors in a differential mode. It is possible to prepare two identical enzyme sensors and either omit or deactivate the enzyme in one of them. This sensor then acts as a reference device. If the calibration curve is constructed by plotting the difference between the two outputs as a function of the concentration of the substrate, the effects of variations in the ambient composition of the sample as well as temperature and light variations can be substantially reduced.

It is sometimes necessary to use two or more enzymes in a *cascade arrangement*. One example is the glucose sensor (Scheme 4-1) in which the overall reaction involves the β-D-glucose oxidase which transfers electrons from glucose to oxygen (i.e., oxidizes glucose), and the catalase (hydrogen peroxidase) which decomposes the by-product of this reaction, the hydrogen peroxide. Catalase is the natural complementary enzyme found with all oxidases, because it protects the biological microenvironment by removing the hydrogen peroxide. An enzymatic cascade arrangement with two or more discrete enzyme layers, each containing its own enzyme, is used. Potentiometric enzyme electrodes described in the literature are listed in Table 4-5.[27]

Despite the large amount of research done on these devices and the relatively detailed understanding of their operation, they have not yet found widespread use. The main reason seems to be the insufficient lifetime (both "shelf" and "in-use") due to the degradation of the *ex vivo* enzyme preparations. Various reversible and irreversible inhibitors reduce their usable lifetime, particularly when they are used in real samples. This

Table 4-5. Potentimetric Enzyme Electrodes[a]

Substrate	Enzyme	Sensor	Stability	Response time	Range (M)
Urea	Urease	NH_4^+ ISE	>4 mo	1–2 min	10^{-2}–10^{-4}
		pH ISE	3 w	5–10 min	10^{-3}–10^{-5}
		Gas (NH_3)	>4 mo	2–4 min	10^{-2}–10^{-5}
		Gas (CO_2)	3 w	1–2 min	10^{-2}–10^{-4}
Glucose	Glucose Oxidase	pH ISE	1 w	5–10 min	10^{-1}–10^{-3}
L-Amino acids	L-AA oxidase	NH_4^+ ISE	>1 mo	1–3 min	10^{-2}–10^{-4}
L-Tyrosine	L-Tyrosine decarboxylase	Gas (CO_2)	3 w	1–2 min	10^{-1}–10^{-4}
L-Glutamine	Glutaminase	NH_4^+ ISE	2 d	1 min	10^{-1}–10^{-4}
L-Asparagine	Asparaginase	NH_4^+ ISE	1 mo	1 min	10^{-2}–10^{-4}
D-Amino acids	D-AA oxidase	NH_4^+ ISE	1 mo	1–3 min	10^{-2}–10^{-4}
Penicillin	Penicillinase	pH ISE	1–2 w	30–120 s	10^{-2}–10^{-4}
Amygdalin	β-Glucosidase	CN-ISE	3 d	10–20 min	10^{-2}–10^{-5}
Nitrite	NO_2 reductase	Gas (NH_3)	3–4 mo	2–3 min	10^{-2}–10^{-4}

[a] Selected from Ref. 27 with permission of Wiley.

problem seems to be mitigated when the enzymes are employed in their "natural" environment, in cell cultures, tissue slices, and even in whole organs.[28] Obviously, such an approach does not lend itself to an exact physical description and to the corresponding mathematical treatment, but it can produce useful practical devices.

4.1.5. Gas Sensors

In potentiometric gas sensors the potential change is obtained from the interaction of electrically neutral gas molecules with the sensor. The sample can be a gas or a liquid phase. Potential difference implies charge separation, so there must be a mechanism which links the primary inter-action of the electrically neutral gas to partitioning of ions or electrons at an interface within the sensor. This is indeed the case in the three types of sensors discussed in this section.

4.1.5.1. Severinghaus Electrodes

Although the original design was made specifically for sensing carbon dioxide,[29] the principle on which this device operates is so general that other gases which fit this scheme will be included in this discussion. These electrodes are based on measurement of local ion-activity variation caused by permeation of gas molecules to the inner electrode compartment and their subsequent hydrolysis (Figure 4-42). The mechanism can be described by a series of equilibrium processes, the first being partitioning of the gas between the sample and the electrode. For the carbon dioxide sensor this step is given by the solubility equilibrium

$$(a_{CO_2})_w = \alpha P_{CO_2} \tag{4-87}$$

where $(a_{CO_2})_w$ is the activity of dissolved CO_2 in water, α the solubility

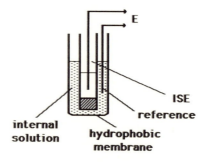

Figure 4-42. Schematic diagram of a Severinghaus electrode.

coefficient, and P_{CO_2} is the partial pressure of CO_2. Hydrolysis inside the electrode proceeds according to the equation

$$CO_2 + H_2O \overset{K_a}{\Longleftrightarrow} HCO_3^- + H^+$$

where K_a is the acid dissociation constant given by the equilibrium expression

$$K_a = a_H a_{HCO_3}/a_{CO_2} \tag{4-88}$$

In principle we have a choice at this point of using a hydrogen ion-selective electrode or a bicarbonate selective electrode to monitor the changes in the activities of these ions. In this particular case an internal pH electrode (or a pH ISFET) is the sensor of choice. The dissociation of one mole of CO_2 yields one mole each of hydrogen and bicarbonate ion, hence equation (4-88) after substitution from equation (4-87) becomes

$$K_a = a_H^2/\alpha P_{CO_2} \tag{4-89}$$

By taking the negative logarithm and rearranging, we then obtain

$$pH = A - 0.5 \log P_{CO_2} \tag{4-90}$$

where $A = 0.5 \, pK_a \log \alpha$. The sensitivity of the Severinghaus electrode can be doubled if the internal filing solution contains a high concentration of bicarbonate (such as 0.1 M $NaHCO_3$). In that case the increase of the bicarbonate ion activity due to dissolved CO_2 is negligible compared to the total concentration, so equation (4-88) becomes

$$K_a' = a_H^+/\alpha P_{CO_2} \tag{4-91}$$

which is then rearranged to

$$pH = A' - \log P_{CO_2} \tag{4-92}$$

Similar sensors for NH_3, SO_2, HCN, etc. can be designed (Table 4-6). It is important to note that if the detected species is hydrogen ion, then *all acidobasic species are mutually interfering.* Improved selectivity can be obtained by judicial choice of the internal potentiometric element and, to some extent, from the differential permeability of the hydrophobic membrane. Thus, e.g., for selective detection of HCN ($pK_a = 3.32$) the internal element should be a CN^--selective potentiometric sensor and the pH of the internal electrolyte should be at least 2 pH units above the pK_a value (e.g.,

Table 4-6. Possible Equilibria Applicable to Gas Electrodes

Diffusing species	Equilibria	Sensing electrode
NH_3	$NH_3 + H_2 = NH_4^+ + OH^-$	H^+
	$xNH_3 + M^{n+} = M(NH_3)_x^{n+}$	$M = Ag^+, Cd^{2+}, Cu^{2+}$
SO_2	$SO_2 + H_2O = H^+ + HSO_3^-$	H^+
NO_2	$2NO_2 + H_2O = NO_3^- + NO_2^- + 2H^+$	H^+, NO_3^-
H_2S	$H_2S + H_2O = HS^- + H^+$	S^{2-}
HCN	$Ag(CN_2)^- = Ag^+ + 2CN^-$	Ag^+
HF	$HF = H^+ + F^-$	F^-
	$FeF_x^{2-x} = FeF_y^{3-y} + (x-y) F^-$	Pt(redox)
HOAc	$HOAc = H^+ + OAc^-$	H^+
Cl_2	$Cl_2 + H_2O = 2H^+ + ClO^- + Cl^-$	H^+, Cl^-
CO_2	$CO_2 + H_2O = H^+ + HCO_3^-$	H^+
X_2	$X_2 + H_2O = 2H^+ + XO^- + X^-$	$X = I^-, Br^-$

pH > 5.5). In that case practically all HCN will be dissociated and there will be no interference from other acidic gases.

The electrical circuit is closed inside the sensor, therefore no external reference electrode is necessary and the Severinghaus-type electrode can be used for measurement in either gaseous or liquid samples. It is important to remember, however, that the potential of the internal reference electrode (Figure 4-6) remains constant. In principle, it would be possible to use a liquid junction but it would add to the complexity of the design. The counterion resulting from the dissociation equilibrium is the only interfering ion and it is present in a very low concentration, so the constancy of the reference potential can be ascertained by careful choice of the internal electrolyte. Thus, for example, in the CO_2 electrode the internal electrolyte contains 0.1 M $NaHCO_3$ and 0.1 M NaCl with the Ag/AgCl internal reference element.

The performance characteristics of several Severinghaus-type electrodes are given in Table 4-7. The general facets are the simplicity of construction and of the operating principle. Since partitioning between the sample and a relatively large internal volume as well as dissolution equilibria are involved, the time constant of the Severinghaus electrode typically 10–20 s. The addition of gas from the outside to the inside of the electrode can be viewed as a titration of the internal volume. For that reason the dynamic range depends on the concentration of the internal electrolyte. Clearly, the lower the concentration the wider the dynamic range. However, for too low a concentration the sensitivity of the electrode (slope) will decrease due to transition from the response given by equation (4-92) to that given by equation (4-90).

Severinghaus electrodes have found wide application in clinical

Table 4-7. Some Specifications of Gas Sensing Electrodes

Species sensed	Sensor	Internal electrolyte	Lower limit (M)	Slope	Sample preparation	Interferences
CO_2	H^+	0.01 M $NaHCO_3$	10^{-5}	+60	<pH 4	
NH_3	H^+	0.01 M NH_4Cl	10^{-6}	−60	>pH 11	Volatile amines
Et_2NH	H^+	0.1 M Et_2NH_2Cl	10^{-5}	−60	>pH 11	NH_3
SO_2	H^+	0.01 M $NaHSO_3$	10^{-6}	+60	HSO_4 buffer	Cl_2, NO_2 must be destroyed (N_2H_4)
	H^+	0.1 M $NaHSO_3$	10^{-4}			
NO_2	H^+	0.02 M $NaNO_2$	5×10^{-7}	+60	Citrate buffer	SO_2 must be destroyed (CrO_4^{2-}) CO_2 interferes
H_2S	S^{2-}	Citrate buffer (pH 5)	10^{-8}	−30	<pH 5	O_2 (ascorbic acid must be added to samples)
HCN	Ag^+	$KAg(CN)_2$	10^{-7}	−120	<pH 7	H_2S (add Pb^{2+})
HF	F^-	1 M H^+	10^{-3}	−60	<pH 2	
HOAc	H^+	0.1 M NaOAc	10^{-3}	+60	<pH 2	
Cl_2	Cl^-	H_2SO_4 buffer	5×10^{-3}	−60	<pH 2	

analysis. It is pertinent to mention here that the general principle of gas permeation through a hydrophobic membrane followed by its detection (with or without its solvolysis) has been used with different types of internal sensor, such as optical, amperometric, conductimetric, or a mass sensor. The choice of internal sensing element depends on the circumstances of the application in which the gas sensor would be used, such as the required time response, selectivity considerations, complexity of instrumentation, and so on.

4.1.5.2. Work Function Sensors

The use of a work function for detection of gases was proposed almost forty years ago. A modestly selective detector for gas chromatography was constructed on the principle of a vibrating capacitor (Kelvin probe). In this device two plates of a parallel-plate capacitor are moved periodically with respect to each other (vibrated). If an electric field exists between the two plates, their movement results in an AC signal which can be zeroed by applying a DC voltage to the plates. This voltage can be related to the surface potential at the two plates.[30,31] This detector has shown promising response characteristics, however, at the same time a much more sensitive and more compact flame ionization detector has been invented and the vibrating capacitor detector for gas chromatography has been all but forgotten. The real reason for the demise of this fundamentally correct idea probably lies in the practical limit on the minimum size of this detector. Nevertheless, as far as we know, it could be regarded as the first predecessor of the devices discussed in this chapter.

The essential element in the operation of the IGFET, including the suspended-gate field-effect transistor (SGFET), is the metal–insulator–semiconductor (MIS) junction. The theory of IGFET is treated in detail in Section 4.1.3.2, so only those features of the MIS junction specific to the operation of SGFET and IGFET will be presented here.

So far, only a few devices employ the chemical modulation of the work function as the source of the primary signal. The palladium gate IGFET for hydrogen sensing is the best example.[32] The energy levels in the palladium–insulator–silicon gate capacitor will be examined first (Figure 4-43). This diagram will also be used to illustrate the physical meaning of the electron work function. In Figure 4-43a the junction is disassembled by a division through the insulator, and both halves are treated as electrically isolated objects. There are three electron work functions to be considered: that of palladium (ϕ_{Pd}), that of an arbitrary metal which does not interact with hydrogen (ϕ_M), and that of silicon (ϕ_{Si}). They are defined in terms of the energy required to remove the electron from the Fermi level of the material and place it just outside the image forces, in the so-called vacuum

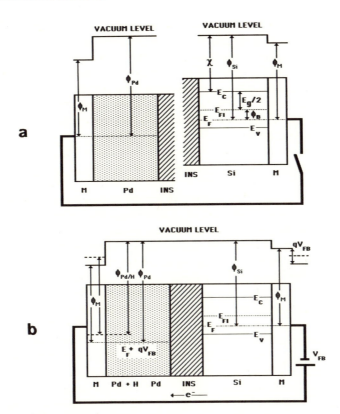

Figure 4-43. Effect of the work function on charge distribution in a metal–insulator–semiconductor junction. (After Ref. 33.)

(reference) level. The two halves are not connected, so their Fermi and vacuum levels are in an arbitrary undefined position with respect to each other. On the other hand, because metal M forms an ohmic junction with palladium (and M with silicon), the Fermi levels in those materials must be equal and the contact potentials $(\phi_{Pd} - \phi_M)/q$ and $(\phi_{Si} - \phi_M)/q$ arise at their interfaces.

It is noteworthy that the surfaces of the palladium and metal are clean, and their work functions are due only to the energy required to remove the electron from the bulk. On the other hand, in the semiconductor the work function has two terms: the electron affinity term χ due to removal of the electron from the bottom edge of the conduction band, and the bulk term $\psi = (\phi_B + E_g/2)$. Furthermore, we specify that there is no electric field and therefore no space charge at the insulator–silicon interface. In other words, the potential difference between the surface and the bulk of the semiconductor is zero ($\phi_{SB} = 0$) and the energy bands are flat. This arbitrary, but convenient, state is called the *flatband condition* (cf. Appendix C). Let us stipulate that throughout the ensuing manipulations the flatband condition will be maintained, if necessary with the help of an externally applied *flatband voltage* V_{FB}. The Fermi level inside the palladium layer is flat by virtue of the high conductivity (no electric field can exist inside a metal at equilibrium). It is further assumed that no trapped charges or oriented dipoles exist inside the insulator or at its interfaces.

We now imagine that the two halves are rejoined (Figure 4-43b). The connecting lead is made again from the same metal M, so that no additional contact potentials develop (they would not affect the final result anyway). The Fermi level in silicon is employed as the starting point and the test charge passed through this structure in a counterclockwise direction; the position of the vacuum level above silicon is defined by ϕ_{Si}. The vacuum level over the insulator is flat (no electric field) owing to the flatband condition. That defines the position of the Fermi level inside the palladium layer which is ϕ_{Pd} below the vacuum level. Since there is an ohmic contact (exchange of electrons) between M and Pd, the Fermi level inside M is the same as in Pd and defines the position of the vacuum level for M (ϕ_M above the Fermi level). The metal M on the left-hand side of Figure 4-43 is the same as on the right, as is the vacuum level. There is a contact potential at that interface, because the Fermi levels in M and in Si are equal. Now it is seen that the vacuum levels in the same metal M are *not equal*. The resulting difference is the flatband voltage V_{FB}, which must be applied externally in order to satisfy the flatband condition. We now go over these steps again, this time adding up the energy contributions beginning and ending at the silican Fermi level so that the total energy is zero:

$$\phi_{Si} - \phi_{Pd} + \phi_M + qV_{FB} + (\phi_{Si} - \phi_M) - \phi_{Si} = 0$$

or

$$qV_{FB} = \phi_{Pd} - \phi_{Si} \qquad (4\text{-}93)$$

Thus, the flatband voltage (multiplied by the test charge) equals the difference in the electron work function of Pd and Si. If there are other charges or dipoles in this structure of whatever origin (nonideal junctions), they must be added to V_{FB}.

There is a different way of looking at the origin of the flatband voltage. The electron work function of palladium is larger (longer arrow) than that of silicon, so it can be argued that electrons are bound more tightly in palladium. When, in our imagination, the two halves of the Pd/INS/Si structure are joined (transition from Figure 4-43a to Figure 4-43b), electrons flow from Si to Pd. Thus the act of joining causes the separation of charge and formation of a potential difference V_{FB}.

Under the flatband condition different metal/insulator/silicon junctions will have different values of V_{FB}. The issue of the chemical modulation of the work function will be discussed in greater detail below. However, in connection with the structures presented in Figure 4-43 let us examine what happens if hydrogen is introduced into palladium at high concentration. As the work function of Pd decreases, the Fermi level is adjusted upward as indicated by the dashed line in Figure 4-43b. Hydrogen does not interact with metal M (as stipulated), so the magnitude of its work function is unaffected but its vacuum level is shifted upward. In order to satisfy the flatband condition the value of V_{FB} must be reduced accordingly relative to its previous value. Such a change in V_{FB} can be measured.

An additional layer of the same metal M has been interposed between Pd and the insulator in the structure shown in Figure 4-44. One could perform the same analysis using the test charge as for Figure 4-43b, but it is unnecessary. It suffices to add the energy contributions in the cycle which begins and ends at the silicon Fermi level (anticlockwise):

$$\phi_{Si} - \phi_M + \phi_{Pd} - (\phi_{Pd} + \phi_M) + qV_{FB} + (\phi_{Si} - \phi_M) - \phi_{Si} = 0 \quad (4\text{-}94)$$

or

$$qV_{FB} = \phi_M - \phi_{Si} \quad (4\text{-}95)$$

The work function of palladium, ϕ_{Pd}, is seen to have disappeared from the flatband voltage, just as ϕ_M disappeared from equation (4-93). This means that despite the modulation of the work function of Pd in this structure, there is no effect on the flatband voltage. The contact potential between Pd and M has changed, but because there are two ohmic junctions on either side of Pd and the change in those potentials cancels out. This situation can be generalized by the statement that: *In order to observe the potentiometric signal due to the chemical modulation of the electron work*

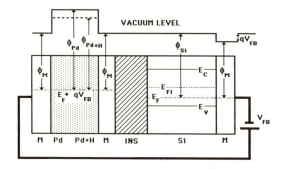

Figure 4-44. Demonstration of the "one-capacitor" rule. (After Ref. 33.)

function of the chemically sensitive layer, at least one interface of this layer must be capacitive or must maintain a constant potential difference. Such a condition is satisfied in the structure shown in Figure 4-43b but not in Figure 4-44. The corollary of this statement is that the layer adjacent to the insulator modulates the response. This rule has one very important implication for chemical sensors. If a conducting layer were to be found which selectively interacts with some species of interest, then such a layer would be highly attractive for chemical sensing. However, it would not be possible to include it in a potentiometric sensor by connecting it to an amplifier (such as an IGFET) with a piece of wire, because such a connection would violate the above rule. Although its work function would be modulated, the effect could not be observed.

In a way the field-effect transistor with Pd gate, investigated extensively by Lundstrom,[32] is the most fundamental CHEMFET. It can be modulated electrically as a normal IGFET and all the equations used to describe the operation of the field-effect transistor (Section 4.1.3.2a) apply. Yet, it responds to a change in the partial pressure of hydrogen. This response is due to modulation of the electron work function of the palladium gate layer as described above.

It has been shown that the mechanism of the response involves a sequence of steps in which the molecular hydrogen dissociates at the Pd surface and diffuses through the Pd layer as atomic hydrogen. It then accumulates at the Pd/SiO_2 interface, where it gives rise to an electric dipole (Figure 4-45). Thus, the primary source of the signal is the increase in the dipole field at this interface. The dissociation, transport, and adsorption equilibria are described by their respective equilibrium constants. For dissociation at the Pd surface

$$H_2 \overset{K_d}{\Longleftrightarrow} 2H_a$$

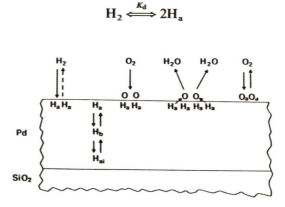

Figure 4-45. Reactions in a Pd/SiO_2 structure. (Reprinted from Ref. 32 with permission of Academic Press.)

for transport from the surface to the bulk

$$H_a \overset{K_b}{\rightleftharpoons} H_b$$

and for adsorption at the Pd/SiO_2 interface

$$H_b \overset{K_i}{\rightleftharpoons} H_{ai}$$

If the Langmuir adsorption isotherm (Table 1-2) applies, then the fraction θ of occupied adsorption sites at the interface satisfies

$$\theta/(1-\theta) = K_b K_i \sqrt{K_d} \sqrt{P_{H_2}} \qquad (4\text{-}96)$$

The shift of the threshold voltage ΔV_T is proportional to the coverage of the interface,

$$\Delta V_T = \Delta V_{Tmax} \theta \qquad (4\text{-}97)$$

Therefore, equations (4-96) and (4-97) yield

$$\Delta V_T = \Delta V_{Tmax} \frac{k_0 \sqrt{P_{H_2}}}{k_0 \sqrt{P_{H_2}} + 1} \qquad (4\text{-}98)$$

where $k_0 = K_b K_i \sqrt{K_d}$.

At high hydrogen pressures, modulation of the bulk term of the work function is also possible. The Pd IGFET operates at elevated temperature (about 150 °C), where the dissociation/diffusion rates for hydrogen are relatively high. The response in air is shown in Figure 4-46.

It is noteworthy that other species which can be catalytically cleaved to yield hydrogen can also be detected by the Pd IGFET. Thus ammonia,

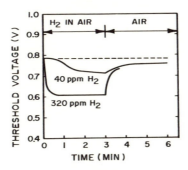

Figure 4-46. Response of a Pd/SiO_2 structure to hydrogen. (According to Ref. 32.)

lower hydrocarbons, and hydrogen sulfide have been found to produce a signal.[32] The general mechanism seems to be catalytic abstraction of hydrogen from these molecules,

$$XH_n + Pd \xrightarrow{K_{dh}} XH_{n-1} + Pd(H)$$

In that case the de-hydrogenation equilibrium constant K_{dh} must be included in k_0.

Hence the Pd layer serves a dual purpose: a catalytic surface which generates the interacting species, and the source of the primary signal. It has been shown that deposition of an additional layer of Al_2O_3 between Pd and SiO_2 further improves the operational characteristics, namely, the drift of these devices.

The scheme in Figure 4-45 also shows the catalytic reaction involving oxygen. If *both* oxygen and hydrogen are present, the steady-state response of the Pd IGFET is described by

$$\Delta V_T = \Delta V_{Tmax} \frac{k_0' \sqrt{P_{H_2}/P_{O_2}}}{k_0' \sqrt{P_{H_2}/P_{O_2}} + 1} \tag{4-99}$$

where $k_0' = K_b K_i \sqrt{c_1/2c_2}$. The rate constants c_1 and c_2 correspond to the rate of dissociation of hydrogen,

$$H_2 \xrightarrow{c_1} 2H_a$$

and the rate of Pd surface catalyzed oxidation,

$$H_2 + \tfrac{1}{2}O_2 \xrightarrow{c_2} 2(OH)_a$$

A limitation of the Pd IGFET device is that its use revolves only around hydrogen and hydrogen-producing species.

Another device operating on the basis of the modulation of the electron work function is an IGFET sensitive to O_2.[34] It incorporates a thin layer of yttria-stabilized zirconia as the solid electrolyte, which is deposited between the gate insulator and a thin (100 Å) Pt electrode. Oxygen penetrates through the (presumably) porous Pt electrode and modulates the work function of the Y_2O_3, ZrO_2 layer. This in turn changes the contact potential at the $Pt/(Y_2O_3, ZrO_2)$ interface. Since the insulator–zirconia interface is capacitive, the device functions as a solid-state potentiometric sensor.

Suspended-gate field-effect structures are more general, because they provide the means by which any chemical species can interact with the electric field within the gate.[33] The arguments developed for the Pd/Ins/Si junction are essentially the same (Figure 4-47). If the chemically selective layer deposited inside the gap at metal M is a semiconductor, then its energy band diagram can be described in terms analogous to the silicon band structure (Figure 4-43). Here, for convenience, the energy band diagram is rotated by 90°. The vacuum level across the insulator is flat, as dictated by the flatband condition. It is assumed that there is no modulation of the work function of silicon. In general this is relatively true for silicon devices such as capacitors or transistors. The second assumption is that there is no adsorption at the (solid) insulator surface. That condition is difficult to realize and verify in practical devices. It can be shown that there is the same fundamental experimental uncertainty involved as in the determination of a single electrode potential or single ion activity. However, for the sake of simplicity we assume that the surface potential of the insulator is constant. According to Figure 4-47 the electron work function of the layer, ϕ_L, is given by

$$\phi_L = \chi_L + \phi_B + E_g/2 \qquad (4\text{-}100)$$

The surface potential χ_L/q (electron affinity by physical definition) will be affected mainly by the adsorption, while the bulk term $\psi = (\phi_B + E_g/2)$ will respond to the modulation of the Fermi level.

The fundamental physical difference between ordinary IGFET and

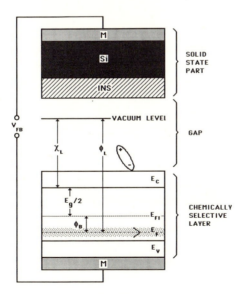

Figure 4-47. Energy relationships in a gap junction. (After Ref. 33.)

SGFET is that, in the latter, there is an additional "insulator" present within the gate structure—the gap (Figure 4-48). From the sensing point of view the most important property of this insulator is to allow the chemical species to penetrate into the transistor gate and interact with the electric field. This condition defines the mode of coupling between the chemically selective layer and the physical part of the sensor. This gap can consist of a vacuum, or a gas, or a nonconducting liquid. For the reasons discussed above the only phase which it *must not contain* is a conductor. Equations describing the operation of a SGFET are formally identical with those governing an IGFET, with two basic differences. First, the electron work function of the gate metal has been replaced by the electron work function of the chemically sensitive layer. This layer can again be metal, such as palladium, in which case it is expected that the SGFET will respond to hydrogen. We will see later that this is indeed so. Second, the changes in the geometry of the gate capacitor affect the transconductance, i.e., the current sensitivity.

    a. Geometrical Effects. A convenient way of discussing this point is to compare the conventional IGFET of the same gate width (W) and gate length (L) with the SGFET. In the SGFET case the gate capacitance, and therefore the drain current, will be smaller by the factor

$$C = C_0 C_g/(C_0 + C_g) \qquad (4\text{-}101)$$

where C_0 is the equivalent IGFET capacitance and C_g the capacitance of the gap. For a very large gap ($C_0 \gg C_g$) the gate capacitance is determined only by C_g which, being low, decreases the current sensitivity of the transistor compared to the equivalent IGFET.

    The drain-current equations of SGFET operated in the nonsaturation and saturation regimes, respectively, are obtained by substituting equa-

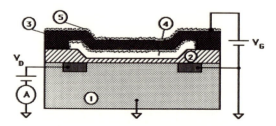

Figure 4-48. Schematic diagram of a suspended gate field-effect transistor: (1) substrate, (2) insulator, (3) suspended metal, (4) gap, (5) selective layer. (After Ref. 33.)

tion (4-101) into the normal drain-to-source current equations for an IGFET. Thus for nonsaturation $(V_D < V_G - V_T)$

$$I_D = \frac{W\mu_n C_0 C_g V_D}{L(C_0 + C_g)}(V_G - V_T - V_D/2) \tag{4-102}$$

and for saturation $(V_D > V_G - V_T)$

$$I_D = \frac{W\mu_n C_0 C_g}{2L(C_0 + C_g)}(V_G - V_T)2 \tag{4-103}$$

The threshold voltage V_T is defined as

$$V_T = (\phi_L - \phi_S) + 2\phi_{FI} - (Q_{SS} + Q_B)/C_0 \tag{4-104}$$

where ϕ_{FI} is the intrinsic Fermi level of the semiconductor and ϕ_S is its work function. For the operation of SGFETs it is assumed that these two terms are constant. The sensing mechanism involves changes in the work function of the chemically sensitive layer, ϕ_L.

The effect of the gap capacitance on the slope of the I_D-V_G curve and on V_T is seen in Figure 4-49. The thickness of the gap in excess of approximately 3000 Å would make this slope (and the signal-to-noise ratio) too low for any practical use. This upper limit on the gap thickness makes the use of thin-layer deposition/etching techniques mandatory for the fabrication of SGFET. On the other hand, if the gap thickness is small $(d \to 0, C_g \gg C_0)$ the gate capacitance approaches [see equation (4-101)] that of the conventional IGFET,

$$\lim_{d \to 0} C = C_0 \tag{4-105}$$

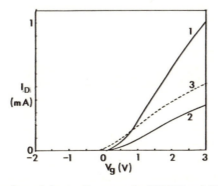

Figure 4-49. Comparison of the I_D-V_G curves for IGFET and SGFET. (After Ref. 33.)

The lower limit of the final gap thickness is given by practical considerations governing mass transport inside the gap during the final stages of deposition of the chemical layer. It is estimated to be 500 Å (at the end of the deposition step) for the suspended-gate area of approximately $10 \times 10 \, \mu m$. Thus, with these two limits, for a SGFET with a 2000-Å-thick solid gate insulator (i.e., SiO_2/Si_3N_4, average relative dielectric constant $\varepsilon = 4$) the drain current is between 15 and 40% of the I_D value of the equivalent IGFET for the equivalent biasing conditions.

The gate geometry affects the threshold voltage V_T through the last term in equation (4-104). The contribution to the electric field arising from the trapped charges in the insulator (Q_{SS}) and from the depletion layer charge (Q_B) depends on the insulator-gate capacitance. The gate is a capacitive divider, so changes in C_g at constant ($Q_{SS} + Q_B$) will also affect V_T. The shift in V_T due to geometrical factors can be observed only in SGFETs with a different gap thickness but otherwise an identical chemically sensitive layer. For different materials this effect is overshadowed by the V_T shift due to the change in ϕ_L. However, once the device has been fabricated this term becomes (to a first approximation) constant, unless the dielectric constant of the gap changes upon exposure to some chemical environment. Such a situation would be possible, for example, in a humidity sensor where the concentration of the water vapor is sufficiently high to cause measurable changes in the bulk dielectric constant of the gap. Obviously, any change in the dielectric constant of the chemical layer would have no effect on V_T as long as that layer is electrically conducting. In any further discussion we shall assume that the changes in V_T due to C in equation (4-104) are negligible compared to changes due to chemically induced variations in the work function term ϕ_L.

b. Thermodynamical Considerations. At equilibrium the number of moles n of all species and their chemical potentials μ in a phase (e.g., in a chemically selective layer) are related through the Gibbs–Duhem equation (A-21) in which the chemical potential μ_i of species i is defined as the molar change in the free energy ∂G [see equation (A-19)]. Thus, if a new species enters the phase the chemical potentials of all species in that phase must change. These include a change in the electrochemical potential of the electron, the Fermi level, and therefore the electron work function, in which case the electrical work is included.

The graphical representation of the situation in the gate of SGFET is shown in Figure 4-47 for a p-type semiconductor, chemically selective layer. The position of the energy level for the dopant, and therefore the position of the Fermi level in the whole phase, depends on the position of the intrinsic Fermi level of the pure material, E_{FI}, on the electron donor/acceptor properties of the dopant, and, if the dopant is charged, on the occupational density of the donor states, E_D. Therefore, for an n-type material the dopant energy level (donor) would be located close to the conduction band edge, E_c, and the Fermi level would be closer to that edge accordingly. In Figure 4-47 the acceptor molecules are assumed to be charged and therefore their distribution depends on their occupancy. This fact is shown by the symbol for a distribution (>) in Figure 4-47.

The electrochemical potential of an electron can be expressed as the sum of electrostatic energy and chemical potential,

$$\tilde{\mu}' = \mu' - e\phi_G \tag{4-106}$$

where ϕ_G is the bulk (Galvani) energy of the phase. Since this energy is given with respect to the vacuum level, it consists of the energy contributions resulting from the bulk potential $\phi = (\phi_B + E_g/2)$ and the surface (dipole) potential χ. Therefore

$$\tilde{\mu}' = \mu' + e\chi + e\psi \tag{4-107}$$

The work function of the chemically sensitive layer ϕ_L then satisfies

$$-\phi_L = \mu_L - e\chi_L \tag{4-108}$$

This latter equation indicates that the chemical modulation of the work function can originate from two effects: action of the guest molecule on the energy state distribution in the bulk of the phase, i.e., by absorption (the term μ_L in the equation), or by modulation of the surface potential χ_L, i.e., by adsorption. Both these terms have a different dependence on the activity of the guest molecule. The chemical potential follows the logarithmic law

$$\mu_L = \mu_L^0 + RT \ln a_i^L \tag{4-109}$$

while the surface concentration depends on the type of adsorption isotherm and usually has the form of a power law. This may, in fact, create some problems in resolving the two types of contribution to the overall change in the work function, because the relative degree of their contribution is not known *a priori*.

c. Chemical Modulation of the Work Function. The issues concerning the relationship between the Fermi level and the electrochemical potential of an electron have been discussed by Reiss.[35] The following section follows his line of reasoning. The conceptual difference between the meaning of the Fermi level, as used by physicists, and the electrochemical potential of an electron, as used by chemists, lies primarily in the reference state. The Fermi level is referenced with respect to the energy of a stationary electron in vacuum, just outside the reach of the image forces (vacuum level). The chemical definition uses the electrochemical potential of an electron in the standard hydrogen electrode as a reference. Furthermore, the Fermi level is defined with respect to one electron (hence "kT"), while the electrochemical potential is defined "per mole" (hence "RT").

Let us consider the number of particles n_l in energy level l having energy ϕ_l and consisting of energy states ω_l:

$$n_l = \frac{\omega_l}{1 + \exp(\phi_l - \mu')/kT} \tag{4-110}$$

This is the so-called normal Fermi–Dirac distribution, in which μ' is the chemical potential of a single particle. When the particle carries charge q, the energy of the individual states is increased by qV, where V is the local electrostatic potential. This defines the so-called electrochemical potential of the particle,

$$\tilde{\mu}' = \mu' + qV \tag{4-111}$$

It is now assumed that the individual energy states are provided by the guest molecules (A), which have entered the phase and can exist in it as electrically neutral (A), or as singly ionized (A^-) or doubly ionized (A^{2-}) species according to the equilibria

$$A + e \Leftrightarrow A^- \qquad \text{and} \qquad A^- + e \Leftrightarrow A^{2-}$$

The total activity of these species is then

$$\omega_l = a_A + a_{A^-} + a_{A^{2-}} \tag{4-112}$$

The normal Fermi–Dirac distribution [equation (4-110)] holds for electrons occupying uncharged molecules A. However, the distribution of electrons through the energy states represented by the ionized host molecules is affected by the electrostatic interactions between these levels. It is described by

$$a_{A^-} = \frac{\omega_l}{1 + \exp[\phi_l(1) - \mu']/kT + \exp[\phi_l(2) - \mu']/kT} \tag{4-113}$$

and

$$a_{A^{2-}} = \frac{\omega_l \exp\{-[\phi_l(2) - \mu']/kT\}}{1 + \exp[\phi_l(1) - \mu']/kT + \exp[\phi_l(2) - \mu']/kT} \tag{4-114}$$

These latter two equations yield the ratio $a_{A^-}/a_{A^{2-}}$ in the form

$$a_{A^-}/aA^{2-} = \exp[\phi_{(2)} - \mu']/kT \qquad (4\text{-}115)$$

Up to this point the statistics of individual particles occupying different energy states has been examined. This is a rather unusual formalism for a chemist, but it is consistent with the definitions given in Figures 4-43, 4-44, and 4-45. If equation (4-115) is now divided by the charge of the electron and the argument of the right-hand side expanded in the Avogadro number, the discussion is transformed into the realm of chemistry and

$$\pi = \pi^0 + RT/F \ln a_{A^-}/a_{A^{2-}} \qquad (4\text{-}116)$$

is derived where π and π^0 are the potential and standard potential of the redox couple A^-/A^{2-}, while the activities are expressed on a molar scale. This is, of course, a classical expression for the redox potential. A similar expression can be derived for the first redox equilibrium using equations (4-110), (4-113), and (4-114).

The purpose of this derivation has been to show that the chemical modulation of the Fermi level, and therefore of the work function, can be described in either statistical or classical terms. The statistical approach allows one to formulate the nature of this effect: if the entering species are electrically neutral and remain electrically neutral inside the solid phase, then the energy distribution of the states within the energy level follows normal Fermi–Dirac statistics. On the other hand, if the molecules are ionized or undergo ionization, then the distribution of energy states depends on their degree of occupation and becomes non-Fermi–Dirac. In either case the effect on the Fermi level is there as long as these states can accept or donate electrons. Thus, all types of electronic interaction involving the guest molecule can be included in this scheme, including complete charge transfer or a partial charge sharing.

From the thermodynamical standpoint, the speciation of the molecules entering the chemically selective layer according to equation (4-112) must be treated as being equivalent to the addition of individual species $(A, A^-, A^{2-}, \text{etc.})$ to that phase. The final position of the Fermi level, and hence the value of the work function of that phase, then depends on the activity (partial pressure) of the species in the gas, on their solubility coefficients in the solid phase, and on the individual dissociation constants K_1 and K_2 in the dissociation equilibrium.

The electrochemical potential of a single phase cannot be measured, but the difference between the electrochemical potentials of two phases can. The classical technique for this measurement is the vibrating capacitor (Kelvin method) mentioned earlier. Measurement with SGFET is conceptually similar, except that nothing needs to be vibrated because the charge density at the capacitor plate is probed by the drain current flowing through that plate parallel to its surface. For geometrical reasons it is evident that the use of the Kelvin probe in a microsensor configuration is not practical.

d. Electrochemical Modification of the Metal Gate. The surface of the suspended gate can be modified with inorganic, organic, or organometallic layers. In principle it is possible to use different deposition procedures. Electrodeposition is particularly attractive owing to the geometry of the gap and its simplicity. The layers are deposited to a thickness of 100–2000 Å depending on the initial thickness of the gap. The deposition is carried out in a narrow space, so it must be performed in such a way that, during electrolysis, the formation of the depletion layer comparable with the thickness of the gap is avoided. This can be achieved by undertaking the deposition in a pulsed mode with individual pulses sufficiently spaced out in time.[33]

The suspended-gate field-effect transistor coated with palladium offers the best opportunity of comparing its performance with the conventional Pd IGFET. The main difference between these two devices is that, in the Pd IGFET, the signal originates from the formation of the dipole at the Pd/insulator interface[32] while in the Pd SGFET it is the surface interaction of hydrogen with palladium which plays the dominating role. Typical responses for exposure to 100 ppm H_2 with nitrogen and air as carrier gases are shown in Figure 4-50.

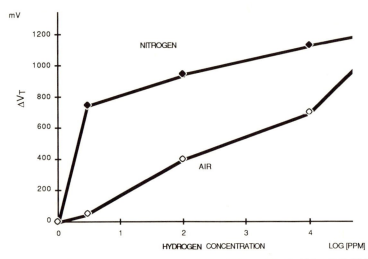

Figure 4-50. Hydrogen response of a Pd SGFET at 140°C. (After Ref. 33.)

Polypyrrole (PP) is an example of an organic, chemically sensitive layer. Its deposition conditions are well known and it is a relatively stable polymer. It is prepared by electrochemical oxidation of pyrrole from different media by a process requiring 2–4 electrons per mole.

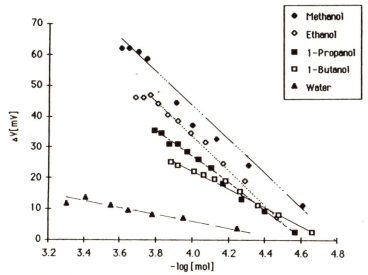

It is a moderately good conductor in its oxidized state ($\sigma =$ 0.01–0.1 S), and can be used as a general matrix and further modified with other compounds in order to change its selectivity. It has been shown that organic compounds can be introduced electrochemically into the bulk of the PP film[36] or that its surface can be derivatized.[37] Furthermore, PP forms an ohmic junction at noble metal electrodes,[38] important for the proper operation of the device. While pyrrole undergoes slow oxidation at room temperature and atmospheric pressure, polypyrrole with a BF_4^- anion is remarkably stable after the initial reaction with air. This means that, if necessary, these devices could be used even at moderately elevated temperatures.

The dependence of the threshold voltage shift on the injected amount of various alcohols (in moles) is shown in Figure 4-51. The fact that the

Figure 4-51. Shift of the threshold voltage in response to alcohol for a SGFET coated with polypyrrole. (After Ref. 33.)

threshold voltage shifts in the positive direction when the transistor is exposed to alcohol vapor means that electrons in polypyrrole become bound more tightly, or that alcohold adsorb on polypyrrole through the $-OH$ group or both. The slope of approximately 60 mV per decade indicates that the bulk term is involved, although such a slope would also be obtained if the adsorption isotherm was logarithmic within this range of concentrations. The primary selectivity of SGFETs with PP layers to alcohols and water may be related to the strong tendency of pyrrole to form hydrogen bonds. Additional chemical selectivity can be achieved by incorporating another functional group, such as nitrotoluenes. The effect of co-deposited nitroarenes on the sensitivity to aromatic compounds is very strong (Figure 4-52): while there is no sensitivity to toluene from PP deposited from methanol, acetonitrile, or acetonitrile/toluene solutions, both 2-nitro- and 4-nitrotoluene PP give a reversible response to benzene and toluene, while PP/3-nitrotoluene yields only a weak, irreversible response.

The results with Pd and with various polypyrrole layers should serve

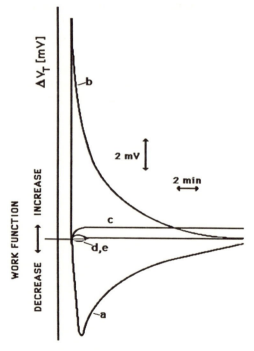

Figure 4-52. Change of threshold voltage in response to injections of 2 μl toluene into a stream of nitrogen. The SGFET was coated with different polypyrrole/nitrotoluene copolymers: (a) PP/4-nitrotoluene, (b) PP/2-nitrotoluene, (c) PP/3-nitrotoluene, (d, e) polypyrrole deposited froth the injection. (After Ref. 33.)

only as examples of possible chemical modifications. There are many more organic, inorganic, and organometallic materials that can be used as chemically selective layers with the SGFET. They can be deposited under different electrochemical conditions and in different combinations, so it is evident that the number of potentially interesting coatings is very large. The main issue then becomes the rational design of chemical selectivity.

There are several other types of device resembling the SGFET physically and in their mode of operation. The first is the transistor with perforated gate.[39,40] For palladium the operation principle is similar to that of the conventional Pd IGFET. The main advantage is the speed of response due to enhanced accessibility of the Pd. For metals which do not allow direct access of the gas between the gate metal and the insulator, another mechanism has been proposed. The perforation, which is produced either photolithographically[40] or by sputtering of the porous metal,[39] allows the gas to access the areas of the insulator in the openings of the gate metal. Atoms or molecules of gases other than hydrogen may adsorb and/or react on the gate surfaces, causing a change in the surface potential of the metal or insulator which affects the channel conductivity in the semiconductor only via the stray capacitance between the metal and the semiconductor. It has been suggested by Lundstrom that, by this mechanism, changes in the potential of any porous conducting layer can be used to alter the field at a semiconductor surface, and hence to detect any species. The main difference between this type of transistor and the SGFET lies in the fact that, in the SGFET, the entire gate region, even underneath the metal, is uniformly accessible to the chemical species and is uniformly modulated by the chemical interactions.

4.1.5.3. High-Temperature Sensors

A potentiometric electrochemical cell consisting of a reference electrode, solid-state electrolyte(s), and an indicator electrode can provide information about the partial pressure of gas in the same way as cells utilizing ion-selective electrodes and liquid electrolytes. The general mechanism is as follows. A sample gas which is part of a *redox couple* permeates into the solid-state structure, usually through the porous electrode, and sets up a reversible potential difference at the interface according to the reaction

$$\text{gas } X \pm e^- \Leftrightarrow (A)^{\pm e}$$

The general cell arrangement is

$$\text{Pt} \left/ \begin{array}{c} \text{constant activity} \\ \text{of species } A^\pm \end{array} \right/ \begin{array}{c} (A^\pm)\text{-ionic} \\ \text{conductor} \end{array} \left/ \begin{array}{c} \text{gas-sensitive} \\ \text{layer AX} \end{array} \right/ \text{Pt, X(gas)}$$

Two requirements must be met in order to obtain a valid measurement. First, the potential difference at one interface (reference) must not be affected by changes in the composition of the sample. This is the usual requirement of a reference potential in any potentiometric measurement. Second, there must be no continuous electronic conduction through the solid electrolyte(s), otherwise the cell would short out internally. The first requirement can be satisfied by providing a constant (reference) partial pressure of gas to one side of the otherwise symmetrical cell. This is the *reference gas electrode* arrangement, which is analogous to a concentration cell. An example of this type is the high-temperature potentiometric oxygen sensor based on yttrium-stabilized zirconia (Figure 4-53). The schematic representation of the cell is

$$O_2(\text{reference}), \text{Pt}/\text{ZrO}_2; \text{Y}_2\text{O}_3/\text{Pt}, O_2 \text{ (sample)}$$

The electrochemical reaction taking place at the two electrodes is identical,

$$\tfrac{1}{2}O_2 + e = \tfrac{1}{2}O_2^-$$

At the operating temperature (100–400 °C) the oxygen ions possess sufficient mobility in the solid electrolyte. The cell voltage E_{cell} is related to the partial pressure of the gas by the Nernst equation for a concentration cell,

$$E_{\text{cell}} = \frac{RT}{F} \ln \frac{P_{O_2}(\text{sample})}{P_{O_2}(\text{ref})} \tag{4-117}$$

The other possibility of realizing constant activity of the potential-determining ion at the reference interface is to choose solid electrolyte in

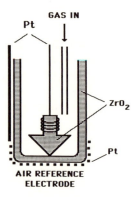

Figure 4-53. Yttria-stabilized zirconia high-temperature potentiometric oxygen sensor with tip-working electrode. (After Ref. 43.)

such a way that the ion of the redox couple is the same as one of the major components of the electrolyte. In that case the change in activity due to the reaction with the gas can be neglected with respect to the overall constant activity of that ion in the salt. This is the *solid-state reference* arrangement. An example is the chlorine sensor (Figure 4-54) in which the reference potential is set up by the constant activity of Cl^- in the solid AgCl electrolyte.[41]

$$Cl_2$$
$$\downarrow$$

$$Ag/AgCl/SrCl_2, KCl/RuO_2$$

reference electrolyte   indicator

$$Ag^+ \quad Cl^- \qquad Cl^- \quad e^-$$

The redox reaction at the indicator electrode (cathode) is

$$Cl_2 + 2e = Cl^-$$

The electrons are supplied from the dissolution of the equivalent amount of silver at the anode (reference) according to

$$Ag = Ag^+ + e^-$$

Upon increase of chlorine partial pressure in the sample, silver ions are generated at the anode and combine with the mobile chloride ions in the AgCl layer. The charge balance is maintained by transport of the chloride ions from the $SrCl_2$, KCl layer. The operation of this sensor is

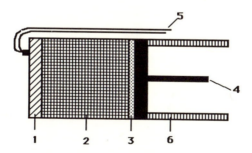

Figure 4-54. High-temperature solid-state ionic sensor for chlorine: (1) RuO_2 working electrode. (2) $SrCl_2$, KCl solid electrode, (3) AgCl reference electrolyte, (4) silver reference electrode and contact, (5) thermocouple (one of the Pt wires provides the connection to the working electrode), (6) Al_2O_3 supporting tube. The glass shield and heater are not shown. (After Ref. 41.)

Table 4-8. Potentiometric Solid-State Gas Sensors[a]

Gas	Cell	Temperature range (°C)	Pressure range	Accuracy $\Delta P/P$ (%)	Time lag
O_2	Rer$\vert$ZrO$_2$–Y$_2$O$_3$$\vert$ME, O$_2$		O$_2$-inert gas: 1–10^{-7} atm		
H_2–H_2O	Ref:Air, Ni–NiO, Pd–PdO, Co–CoO	500–800	CO–CO$_2$, H$_2$–H$_2$O; 10^{-8}	2–5	0.1 s–a few s
CO–CO_2	ME:Pt, Ag		10^{-27} atm		
Cl_2	Ag$\vert$SrCL$_2$–KCl–AgCl$\vert\vert$ME, Pt, Cl$_2$ Ref:Ag$\vert$Ag$^+$ ME:graphite, RuO$_2$	100–450	10^{-6}–1 atm	<5	<1 min
SO_2, SO_3	Ref$\vert$K$_2$SO$_4$$\vert$ME, SO$_2$+SO$_3$+O$_2$ Ag$\vert$K$_2$SO$_4$–Ag$_2$SO$_4$$\vert$ME, SO$_2$+SO$_3$+O$_2$ Air, Pt$\vert$ZrO$_2$–CaOK$_2$SO$_4$$\vert$ME, SO$_3$, air Ref:Ag$\vert$Ag$^+$ ME:Pt	700–900	>10^{-6} atm	3	a few s–a few min
	Ag$\vert$Li$_2$SO$_4$–Ag$_2$SO$_4$$\vert$Pt, SO$_2$+SO$_3$+air Pt, SO$_2$+SO$_3$+O$_2$$\vert$Na$_2SO_4$$\vert$Pt, SO$_2$+SO$_3$+O$_2$	500–750 700	10^{-5}–10^{-2} atm in air >10^{-5} atm	7	$t_{98} = 8$ min
H_2	Ref$\vert$H.U.P.$\vert$ME, H$_2$ Ref:Pd or Pt–H$_2$, PdH$_x$ ME:Pd or Pt, moist atmosphere Pd$\vert$β–β″Al$_2$O$_3$(Na)$\vert$Pt, N$_2$+H$_2$	20	10^{-4}–10^{-1} atm	15 30	$t_{90} \sim 5$ s 1–3 min

Species	Cell	Temperature (°C)		Measuring range	t_{90}
CO	$O_2 + CO$, $MR\|ZrO_2-Y_2O_3\|ME$, $O_2 + CO$ $MR:Al_2O_3-Pt$ $ME:Pt$	250–350		$0-5 \times 10^{-4}$ atm in air	~ 5 min
S_x	$Ag\|Ag\|\|Ag_2S$, S(vap) $Ag\|\beta-Al_2O_3(Ag)\|Ag_2S$, S (vap) $Ref[CaS-Y_2S_3]Pt$, S_x	90–400 90–800 600–900	50	 $<10^{-6}$ atm	a few s
H_2-H_2S	$Ref[CaF_2-CaS]Pt$, $H_2 + H_2S$	700–950	5	$3 \times 10^{-3} < \text{ratio} < 0.2$	2.5 h at 950 °C
CO_2	$Ag\|K_2CO_3-Ag_2SO_4\|Pt$, CO_2 $Ref:Ag\|Ag^+$	700–800		$>10^{-6}$ atm	a few s–a few min
NO_2	$Ag\|Ba(NO_3)_2-AgCl\|Pt$, NO_2 $Ref:Ag\|Ag^+$	500		$>10^{-6}$ atm	a few s–a few min
I_2	$Ag\|KAg_4I_5\|Pt$, I_2	40	3	$>10^{-7}$ atm	a few s–a few min
Na	$Na(vap)\|\beta-Al_2O_3(Na)\|Na$ (vap)	200–360		$10^{-10}-10^{-5}$ atm	10 min

[a] ME and MR denote measuring electrode and reference electrode, respectively. (Reprinted from Ref. 43 with permission of Pergamon Press.)

predicated by the low solubility of Cl_2 in the AgCl phase. The operating temperature range of this sensor is 100–400 °C. The dynamic range is $1-10^{-6}$ atm P_{Cl_2} and the response time is in minutes.

High-temperature potentiometric sensors for other gases, such as O_2, Cl_2, CO/CO_2, SO_2/SO_3, NO_2, I_2, Na, H_2S, and H_2, have been described (Table 4-8). Their principle of operation is the same, but their construction method is largely determined by the conditions existing in the application. Thus, for example, the oxygen zirconia sensor for measurement of O_2 in molten steel (1600 °C) has to be designed in such a way that the thermal expansion coefficients of the different layers in this device are matched. On the other hand, the room-temperature potentiometric oxygen sensor can be constructed[42] by using another set of materials:

$$O_2$$

$$Pt/Sn/Sn, SnF_2/LaF_3/Pt \text{ (black)}/Pt$$

$$e^- \quad Sn^{2+} \quad F^- \quad OH^- \quad e^-$$

In this case the reversible reduction of oxygen takes place at the LaF_3/Pt (black) interface treated with water vapor. It is known that OH^- and F^- ions can move in LaF_3. Thus, the changes in the concentration of the hydroxyl ions which are part of the redox couple are compensated by the transport of the F^- ions across the Sn, SnF_2/LaF_3 boundary in order to maintain the electroneutrality in the LaF_3 crystal. Oxidation of Sn resulting in the formation of SnF_2 at the Pt/Sn, SnF_2 reference electrode completes the cell reactions. Clearly, the reactions involved in this scheme would not be applicable in aqueous solutions, owing to the hydrolysis equilibria involving most of the species. However, apart from this fact the underlying principles of these sensors and of ion-selective electrodes are essentially the same.

The key information necessary for the rational design and development of high-temperature solid electrolyte sensors is knowledge of the thermodynamical (phase diagrams), mechanical, and electrochemical properties of the materials used in their construction.[43,44] These are equilibrium sensors, so the gases detectable are largely limited by the temperature of their operation. The domain of application comprises inorganic gases whose solubility in the given material of the sensor is usually sharply limited. Therefore, the issue of selectivity with high-temperature potentiometric sensors is less severe than with most other sensors.

4.2. Amperometric Sensors

In this group of electrochemical sensors the information is obtained from the current–concentration relationship. The two most important issues to be discussed are, first, the origin of the signal for various types of amperometric sensor and, second, the selectivity. The domain of amperometric sensors from the electrochemical point of view is shown in Figure 4-1. The indicator electrode is either a cathode or an anode, depending on whether the electrons are added to or withdrawn from the sample. Although the chemical changes that take place at the electrodes are different in these two processes, the principle of obtaining the sensor signal is not. The general condition governing the closed circuit must again be satisfied, but the requirements on the stability of the reference electrode are much less critical when compared to the potentiometric sensors (Section 4.1.1.5).

The chemical transformation which takes place at the electrodes during the passage of current is governed by Faraday's law [equation (4-3)], by the mass transport equations whose form depends on both the geometry of the electrode and the experimental arrangement, and by the current–voltage equation (4-4). Questions concerning the quantitative nature of the amperometric signal and its selectivity are closely related. They can be viewed from the perspective of the equivalent electrical circuit representing the amperometric sensor experiment in which the electrical current passes through the sensor and encounters several resistances in series (Figure 4-55): the charge-transfer resistance $R_{ct}(W)$ of the working electrode and the charge-transfer resistance R_{ct} (aux) of the auxiliary electrode, where $j_{0,W} = i_{0,W}/A$

$$R_{ct}(W) = RT/nF j_{0,W} A_W \qquad (4-14)$$

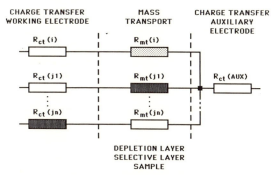

Figure 4-55. General equivalent electrical circuit for an amperometric sensor. The relative values of the resistances determine the path of the sensor current, i.e., selectivity. In this case dark shaded resistances are highest and block the current, i.e., blockage of charge transfer for species j_n and blockage of mass transport for species j_1. The determinant i itself is mass transport limited $[R_{mt}(i) \geqslant R_{ct}(i)]$.

The resistance to mass transport through the chemically selective layer, various membranes, and the sample itself will be labeled $R_{mt}(1)$, $R_{mt}(2)$, etc., respectively, where

$$R_{mt1} = \eta_{mt1}/nF[AD(dC/dx)]_1 \tag{4-118}$$

These resistances are arranged in series, so different voltage drops η will develop across them. They are called *charge-transfer overpotential* (η_{ct}) and *mass-transport overpotential* (η_{mt}), respectively. In an amperometric experiment the auxiliary electrode is always much larger than the working electrode, i.e., its charge-transfer resistance is negligible and need no longer be considered. We know that the total current is the sum of contributions from all species which are electroactive for the given potential applied at the working electrode. In order to achieve selectivity, it is necessary to block currents due to interferants by making their resistances high compared to those of the determinand. The amperometric sensors to be discussed in this section employ different means of achieving this goal, ranging from selection of the operating potential by the catalytic properties of biological layers to controlled mass transport through perm-selective membranes. In terms of Figure 4-55, we can say that if the charge-transfer resistance for interferant j is greater than the corresponding charge-transfer resistance for the determinand $[R_{ct}(j) \gg R_{ct}(i)]$, then the selectivity is controlled by the electrode kinetics. On the other hand, if $R_{ct}(j) \sim R_{ct}(i)$ and $R_{mt}(j) \gg R_{mt}(i)$ then the selectivity will be under mass-transport control.

We have already seen that the electrode and solution form a double-layer capacitor. The minimum energy of this capacitor occurs at the point of zero charge when the electric field between the inner Helmholtz plane and the electrode is at minimum. As the energy of this capacitor is increased by applying the potential difference to this interface, a point is reached when electrons are transferred between the solution and the electrode and electrolysis takes place. Let us assume that a mixture of compounds $(Ox_1, Ox_2,...)$ in their oxidized form are present in the solution and that they are electroactive at different applied potentials. Their ability to accept electrons from the electrode is characterized by their values of the standard (or half-wave) potential. The more positive the value of E_O, the more easily will they be reduced. As the potential of the electrode is made increasingly more negative, the reduction of Ox_1 takes place first, followed by that of O_2, and so on. Thus at potential E_1 only Ox_1 will be reduced. At the more negative potential E_2 reduction of *both* Ox_1 and Ox_2 will take place. In order to determine these two species a measurement will have to be made at two potentials. Such a measurement should be conducted in the diffusion limiting-current region, so for 0.1% resolution the two half-wave potentials must be at least 180 mV apart (assuming they are both fast reactions). Given an approximately 2.5 V width of the electrochemical window, the maximum number of electroactive species that can be fitted into this scheme is approximately 13. This also assumes that their half-wave potentials would be equally spaced and that they would all be reversible (fast). In reality, the number of different species which can be realistically determined in a mixture is at the most four to six. Thus, the choice of potential offers only a very limited selectivity.

Modulation of $R_{ct}(W)$ values for individual interfering species and for the determinand is a desirable and powerful tool in the design of amperometric-sensor selectivity. However, it would be unwise to base the quantitative concentration–current relationship on such modulation because it is an exponential function of the applied potential [see equation (4-4)]. In other words, kinetic control of the current is too sensitive to the conditions of the electrode, solution resistance, potential of the reference electrode, temperature, etc. to be a sole source of selectivity in a reproducible and stable sensor. It must be combined with the selectivity derived from the mass-transport resistances.

There are three factors affecting the transport of chemical species: diffusion, migration, and convection, as given by the Nernst–Planck equation for transport in the x-direction,

$$J_i = -D_i \, dC_i/dx - z_i FU_i C_i \, d\phi/dx + C_i(v_x) \tag{4-119}$$

where v_x is the velocity of the solution in the x-direction; the other terms have been explained in Section 4.1.1.3. In most cases the measurement is conducted in the presence of the excess of the inert electrolyte. This means that the second term (migration) can usually be neglected. Nevertheless, even if it contributes significantly to the signal, it is constant for a constant electric field at the given potential and, as such, it does not usually present any problems from the sensing point of view. On the other hand, convective contribution to the mass transport can create problems because the total contribution to the flux J_i in equation (4-119) (or to R_{mt}) depends on v_x. Therefore, the amperometric sensor should be designed such that v_x is constant, in which case only the mass-transport resistance term controls the signal. Obviously, constancy of the v_x term can be achieved in various controlled flow arrangements (e.g., FIA) or with electrodes which move with respect to the sample in some controlled manner (e.g., rotating or vibrating electrodes). Such operating conditions clearly impose additional constraints and may impair the convenient use of the sensor. They are usually more suitable for applications in sensor systems. In summary, it is desirable that the current be limited only by the diffusion and independent of the relative motion of the sensor with respect to the sample.

In the current–voltage equation (4-4) the concentrations $C(0, t)$ of the oxidized and reduced species at the surface of the electrode are not known explicitly. However, as the difference between the applied and the standard potential increases, the rate of reduction (or oxidation) becomes higher and higher until finally the potential is reached at which all electroactive species that can reach the electrode surface are reduced (or oxidized) and their surface concentration becomes effectively zero. Thus, the current becomes *diffusion limited* [i.e., limited by the value of $R_{mt}(i)$].

This situation is now examined for the case of an electrochemical reaction between the oxidized (O) and reduced (R) forms of a fast redox couple when both O and R are soluble and only O is initially present in the solution,

$$O + ne \Leftrightarrow R$$

The reduction current is proportional to the area of the electrode A, the number of electrons n, and the difference between the bulk concentration C_O^* and the surface concentration $C_O(0, t)$,

$$i = nFAm_O[C_O^* - C_O(0, t)] \qquad (4\text{-}120)$$

The mass transport coefficient m_O is regarded as constant. As the value of E becomes more negative, the current is limited by the *mass transport*, i.e., every molecule of O which reaches the surface of the electrode is immediately reduced to R, therefore

$$\lim_{E \to -\infty} C(0, t) = 0 \qquad (4\text{-}121)$$

and the so-called (mass-transport) *limiting* current i_L is reached (Figure 4-1). This holds for both kinetically slow and fast electron transfer reactions, as can be seen from Figure 4-3. Thus an analytically important relationship is obtained and shows the limiting current to be a linear function of the bulk concentration of the determinand O,

$$i_L = nFAm_O C_O^* \qquad (4\text{-}122)$$

The reduced form R is produced at the electrode surface, from which it is transported to the bulk of the solution where its concentration is zero,

$$i = nFA \, m_R \, C_R(0, t) \qquad (4\text{-}123)$$

quantity m_R being the mass-transport coefficient for the reduced species. As specified above, the electrochemical reaction is very fast, which means that the Nernst equation must be satisfied for all values of the surface concentrations of O and R. Thus

$$E = E^0 + \frac{RT}{nF} \ln \frac{C_O(0, t) f_O}{C_R(0, t) f_R} \qquad (4\text{-}124)$$

When equations (4-120), (4-122), and (4-123) are combined with equation (4-124), the voltammetric equation is obtained in the form

$$E = E^0 + \frac{RT}{nF} \ln \frac{i_L - i}{i} + \frac{RT}{nF} \ln \frac{m_R f_O}{m_O f_R} \qquad (4\text{-}125)$$

If the third term on the right-hand side is constant it can be included in the standard potential, in which case

$$E = E^{0'} + \frac{RT}{nF} \ln \frac{i_L - i}{i} \qquad (4\text{-}126)$$

For a current $i = i_L/2$, the logarithmic term drops out and the so-called half-wave potential $E_{1/2}$ equals approximately the standard potential $E^{0'}$.

Expressions similar to equation (4-126) can be obtained for the case where both O and R are initially present in the solution,

$$E = E^{0'} + \frac{RT}{nF} \ln \frac{i_{L,c} - i}{i - i_{L,a}} \qquad (4\text{-}127)$$

with

$$E^{0'} = E^0 + \frac{RT}{nF} \ln \frac{m_R f_O}{m_O f_R} \qquad (4\text{-}128)$$

For $i = (i_{L,c} + i_{L,a})/2$, $E_{1/2} = E^{0'}$.

If only R is initially present,

$$E = E^{0'} + \frac{RT}{nF} \ln \frac{i}{i - i_{L,a}} \qquad (4\text{-}129)$$

with

$$E^{0'} = E^0 + \frac{RT}{nF} \ln \frac{m_R}{f_R} \qquad (4\text{-}130)$$

For sensing purposes we wish to apply a constant potential to the working electrode somewhere in the region of the diffusion limiting-current plateau, and follow the current as a function of concentration according to equation (4-122). For that relationship to hold the value of the mass-transport coefficient m_O must remain constant during the measurement. It will be seen that this is true only for special conditions.

If the solution moves with respect to the electrode, then there is a stagnant layer (Prandtl layer) formed close to the surface. In a first approximation it is assumed that the conditions of linear diffusion apply over this layer. This is the basis of the Nernst diffusion-layer model. This model assumes that the Nernst diffusion layer coincides with the Prandtl layer, in which case the mass-transport coefficient for species O is expressed as

$$m_O = D_O/\delta_O \tag{4-131}$$

and the current is given by equation (4-120) in the form

$$i = nFAD_O[C_O^* - C_O(0, t)]/\delta_O \tag{4-132}$$

The thickness of the stagnant layer is inversely proportional to the tangential velocity of the solution near the electrode. Therefore the qualitative observation that quantity i_L increases with increased stirring rate of the sample is easily understandable.

If the Nernst diffusion-layer model is applied to an unstirred solution, we must expect the passage of current to cause the formation of a *depletion layer* whose thickness δ_O will grow with time. In time this layer will extend from the electrode surface to the bulk of the solution over hundreds of angstroms. The time dependence of δ_O can be estimated by the Einstein formula for linear diffusion,

$$\delta_O \cong 2(Dt)^{1/2} \tag{4-133}$$

and substituted into equation (4-131) to yield

$$m_O \cong D_O^{1/2}/2t^{1/2} \tag{4-134}$$

It is thus seen that under quiet (unstirred) electrochemical conditions and under conditions of linear diffusion [equation (4-133)] the value of the mass-transport coefficient decreases with the square root of time. If the potential in the diffusion limiting-current region is applied as a single step to the electrode [equation (4-122)], then the diffusion limiting current also decreases with time,

$$i_L(t) \cong \frac{nFA \, D_O^{1/2} C_O^*}{2t^{1/2}} \tag{4-135}$$

The approximate nature of this latter equation is the consequence of using the approximate Einstein relationship, equation (4-133), in its derivation. Nevertheless, it expresses the important characteristics of the experiment carried out under *chronoamperometric* conditions.

The exact solution for the time dependence of the current at a planar electrode imbedded in an infinitely large planar insulator, the so-called semi-infinite linear diffusion condition, is obtained by solving the diffusion equation under the proper set of boundary and intial conditions.[4] This results in the time-dependent concentration profile

$$C_O(x, t) = C_O^* \{1 - \text{erf}[x/(4D_O t)^{1/2}]\} \qquad (4\text{-}136)$$

whose derivative with respect to x yields the concentration gradient at the surface of the electrode $(x = 0)$,

$$dC(0, t)/dx = C_O^*/(\pi D_O t)^{1/2} \qquad (4\text{-}137)$$

When substituted into Fick's law, equation (4-137) becomes the Cottrell equation, which exactly describes the variation in the diffusion limiting current with time,

$$i_L(t) = nFA \, D_O^{1/2} C_O^* / \pi^{1/2} t^{1/2} \qquad (4\text{-}138)$$

The concentration profiles for semi-infinite planar diffusion are shown in Figure 4-56 below. The difference between the exact [equation (4-138)] and approximate [equation (4-135)] solutions is seen to be only 11%.

4.2.1. Microelectrodes

From the sensing point of view the time variation of the signal at constant concentration is clearly unacceptable. Only a slightly better situation exists under conditions of controlled motion of the sample and is practical mostly in sensor systems. Fortunately, it has been realized that for some microelectrodes with smallest dimension on the order of a few microns the current attains the time-independent steady-state value (Figure 4-56)

$$i_L(t) = \frac{nFA \, D_O^{1/2} C_O^*}{\pi^{1/2} t^{1/2}} + \frac{nFA \, D_O C_O^*}{r_0} \qquad (4\text{-}139)$$

where r_0 is the radius of the hemispherical electrode and A equals $4\pi r_0^2$ for a sphere and $2\pi r_0^2$ for a hemisphere. The first term on the right-hand side is the Cottrell term. As time progresses the first term becomes negligible compared to the second and the current reaches its steady state (Figure 4-56). The conditions during which this happens involve a 1% error,

$$Dt/r_0^2 = 3000 \qquad (4\text{-}140)$$

Thus, for a typical diffusion coefficient $D = 10^{-5}\,\text{cm}^2\,\text{s}^{-1}$ the 99% steady state is reached within 3 s for a 1-μm-diameter electrode.

Even more important is the shape and depth of the concentration profiles which characterize the thickness of the depletion layer for microelectrodes (Figure 4-57). The exact solution yields

$$C_O(r, t) = C_O^*\{1 - (r_0/r)\,\text{erf}[(r - r_0)/4D_O t]^{1/2}\} \qquad (4\text{-}141)$$

The most important difference between the spherical microelectrode and the planar macroelectrode is that the concentration profile for the former depends on its geometry, being shallower for smaller radius, while for the planar macroelectrode it is not so. As expected, initially the profiles are similar but as time approaches the steady-state limit $(t \to \infty)$ the concentration profile is given by

$$C_O(r, t \to \infty) = C_O^*[1 - (r_0/r)] \qquad (4\text{-}142)$$

The explanation for this behavior rests on the fact that in a short time the edge of the depletion layer is initially close to the electrode surface, which looks like a parallel plate. As time progresses the area through which the molecules enter the depletion layer becomes larger and larger. In other words, the molecules are supplied to the electrode surface from a solid angle whose volume increases with distance from the electrode and therefore with time of electrolysis. The shallow nature of the concentration

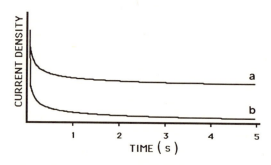

Figure 4-56. Normalized current response to the potential step calculated from equation (4-139) for electrodes of radi (a) $r = 0.5$ cm and (b) $r = 5\,\mu$m. The diffusion coefficient D is ataken as $10^{-5}\,\text{cm}^2\,\text{s}^{-1}$, the concentration C is 10^{-3} M, and $n = 1$.

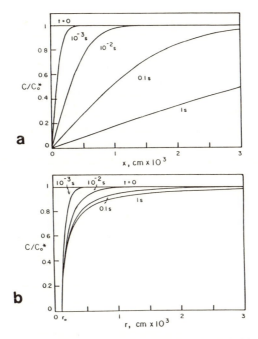

Figure 4-57. Depletion layer at (a) a planar macroelectrode and (b) a spherical micro-electrode. (Reprinted from Ref. 45 with permission of Marcel Dekker.)

profile at the spherical microelectrodes is significant from the sensing point of view. The sensitivity of the current to the sample motion can be seen by changing the thickness of the depletion layer and therefore the mass-transport coefficient. It is much easier to confine the shallow depletion layer of the microelectrode to an additional layer placed at the surface of the microelectrode than at the macroelectrode. Thus, the mass-transport coefficient becomes defined effectively by the thickness of this additional layer and the response of the sensor is largely undisturbed by the motion of the medium. There are other advantages of microelectrodes as compared to macroelectrodes. The current is small (nA to pA), hence the voltage drop due to the electrical resistance of the medium is negligible. For this reason one can perform electrochemical experiments in media which would be unsuitable for macroelectrodes, such as hydrocarbon solvents and frozen electrolytes. Also, the double-layer capacitance is very small because the area of the electrode is small. This means that very fast modulation experiments can be performed. Thus, electrode experiments with a 30-μm-radius disk have been conducted in the nanosecond range. For the same reason the charging current on microelectrodes is much smaller, which improves the signal-to-noise ratio in experiments where the applied voltage is modulated. The small physical size of these electrodes is, of course, an important advantage also for *in vivo* experiments and for measurements in microliter-size samples.

The geometry of the microelectrodes is critically important from both the viewpoint of the mathematical treatment and their performance. The diffusion equations for spherical microelectrodes can therefore be solved exactly, because the radial coordinates for this electrode reduce to a point. On the other hand, microelectrodes with any other geometry investigated until now do not possess a closed solution. One would expect a microdisk electrode, which is easier to make, to behave identically to a microsphere electrode. This is not so, because the center of the disk is less accessible to electroactive species than its periphery. Hence a nonuniform current density exists at this electrode. Its current behavior is divided into a short time and a steady-state region. For the latter ($t \to \infty$) the current is

$$i_L(t) = 4nFD_O A C_O^*(1 + 0.72\, t^{-1/2} + 1.22\, t^{-3/2} + \cdots) \qquad (4\text{-}143)$$

Steady-state currents at band, cylindrical, and ring electrodes as well as on electrode arrays have been investigated; the results can be found in the specialized literature. Differences in the behavior of these electrodes are significant and should be considered in the design of special sensor applications. The common feature of all geometries is that the dimensions of the Nernst diffusion layer are comparable with the smallest dimension of the electrode. This is particularly important in microelectrode arrays, where the closely spaced electrodes at steady state will behave as a single large elec-

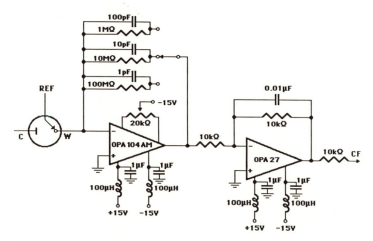

Figure 4-58. Diagram of a picoammeter current booster suitable for use with any conventional potentiostat for measurements with microelectrode. Operational amplifiers are Burr–Brown. (Reprinted from Ref. 47 with permission of the American Chemical Society.)

trode whose active area is given by the sum of the microelectrodes *and* the insulator in between. Obviously, this kind of behavior will be obtained only at times when the Nernst diffusion layers from individual electrodes begin to overlap. A discussion of these effects is beyond the scope of this book and the reader is referred elsewhere for more details.[45,46]

Although potentiometric electrodes of micron and submicron dimensions have been used by electrophysiologists for a long time, amperometric microelectrodes have found comprehensive application only after high-sensitivity current amplifiers became routinely available. A circuit diagram of one such simple high-sensitivity current booster[47] is shown in Figure 4-58. The voltage drop due to the solution resistance is not a problem at low currents, so it is unnecessary to use three-electrode potentiostats.

4.2.1.1. Amperometric Biosensors

The most effective means of achieving chemical selectivity is to modify the electrode with either a synthetic or an enzyme-containing layer. Amperometric sensors that employ enzyme layers will be discussed first. The primary function of the enzyme is to affect the conversion of the electrochemically inactive substrate (determinand) to some electroactive species which can be monitored amperometrically. Thus, the general mode of coupling is based on the fact that some electroactive species is either generated or destroyed by the enzymatic reaction. This idea can be represented as

$$S + E \underset{k_{-1}}{\overset{k_1}{\rightleftharpoons}} ES \overset{k_2}{\Longrightarrow} P + E \qquad \text{and} \qquad P \pm ne \overset{E_1}{\rightleftharpoons} Q_1$$

In principle there is no requirement on how fast the electrochemical reaction should be because, as we have seen, it is usually possible to apply the working potential E_1 at which the kinetics of the charge transfer are fast.

In Figure 1-6 the boundary between the enzyme-containing layer and the transducer was assumed to have either a zero flux or a finite flux of chemical species. The amperometric enzyme sensors stand apart from other types of chemical enzymatic sensor using a different kind of transducer precisely on this distinction. Although the enzyme kinetics are described by the same Michaelis–Menten scheme and by the same set of partial differential equations, the boundary and initial conditions differ if one or more of the participating species move through this boundary. The general diffusion-reaction equation applies to every species in exactly the same manner as was discussed in Section 1.2.2. Although many amperometric enzyme sensors were built in the past by adding the enzyme layer to macro-electrodes, enzyme sensors based on microelectrode geometry offer many advantages, which have their origin in the properties of microelectrodes as discussed in Section 4.2.1.

Many complicated enzymatic schemes, including several enzymes in one layer or multiple enzymatic layers in series, can be used. If co-substrates are involved it is also possible to oxidize/reduce one of the co-substrates in order to obtain information about the determinand. There are simply too many different possibilities with this type of sensor.

There are two main ways in which the enzyme can be employed in amperometric sensors. It can be immobilized within a thick gel layer, in which case the depletion region due to the electrochemical reaction is confined within this layer. This arrangement is similar to that used in any other type of enzyme sensor. Alternatively, the enzyme can be immobilized at the surface of the electrode or even within the electrode material itself, in which case the depletion region extends into the solution just as it would for an unmodified electrode. The enzyme can then be viewed only as an "electrocatalyst" facilitating electron transfer, which would otherwise have too low a rate constant for the charge transfer to occur. The general principles of the design and operation of these two types will be illustrated by the example of the most studied enzymatic sensor, the glucose electrode (see Scheme 4-1, Section 4.1.4).

There are several species in this reaction that can be used for electrochemical sensing. The proton released from the gluconic acid was employed in the potentiometric glucose electrode; the amperometric sensor can be based on oxidation of hydrogen peroxide, on reduction of oxygen, or on oxidation of the reduced form of glucose oxidase itself. The principles governing amperometric enzyme sensors will be demonstrated with the aid of these three modes of operation.

a. Oxidation of Hydrogen Peroxide. For a cylindrical enzyme microelectrode the diffusion-reaction equation can be expressed in spherical coordinates,

$$\frac{\partial C}{\partial t} = D\frac{\partial^2 C}{\partial r^2} \pm \frac{V_m C}{\Re(pH)(C + K_m)} \tag{4-144}$$

The sign in front of the reaction term is positive only for hydrogen peroxide. Also, function $\Re(pH)$ can be made constant by operating the sensor in a medium of high buffer capacity. This is clearly a distinct advantage when compared to potentiometric sensors in which the buffer capacity represented a major interference.

The initial and boundary conditions again depend on the model and on the operating conditions. Initially (at $t = 0$) there is no glucose (G) or hydrogen peroxide (H) present in the gel, which is saturated only with an equilibrium concentration of oxygen (O):

$$C_G(r, 0) = 0 \qquad \text{for} \quad 0 < r < d$$

$$C_H(r, 0) = 0 \qquad\qquad\qquad (4\text{-}145)$$

$$C_O(r, 0) = \alpha_O C_O^*$$

where α_O is the partitioning coefficient for oxygen between the solution and the gel. After the reaction starts ($t > 0$) the boundary conditions are given by the following conditions: there is no glucose present at the surface of the electrode; the reaction takes place entirely within the gel layer,

$$C_G(0, t) = C_G^*(\delta C_G/\delta r)(L, t) = 0 \qquad (4\text{-}146)$$

The oxygen consumed by the primary glucose oxidation is completely regenerated by the oxidation of hydrogen peroxide at the surface of the electrode. Therefore the reaction is never limited by the concentration of oxygen,

$$C_O(0, t) = C_O^*(\delta C_O/\delta r)(L, t) = -(D_H/D_O)(\delta C_H/\delta r)(L, t) \quad (4\text{-}147)$$

The potential of the electrode is set in the diffusion limiting plateau region of hydrogen peroxide. Therefore the concentration of hydrogen peroxide at the surface of the electrode is always zero. Furthermore, there is no hydrogen peroxide in the bulk of the solution which is stirred. Therefore its concentration at the gel–solution boundary is also zero,

$$C_H(0, t) = 0 \qquad \text{and} \qquad C_H(L, t) = 0 \qquad (4\text{-}148)$$

Equation (4-144) can be solved exactly[48] for the above initial [equation (4-145)] and boundary [equation (4-146)] conditions to yield the explicit solution for the concentration profiles within the enzyme layer (Figure 4-59). This equation is long and complicated, however, its advantage is that it can be programmed on a small calculator and used for the design and verification of individual sensors. The current is obtained by evaluating the concentration gradient of the hydrogen peroxide at the surface of the electrode, substituting it into Fick's law, and multiplying it by the charge:

$$i_L = -2FAD_H(\delta C_H/\delta r)(0, t) \qquad (4\text{-}149)$$

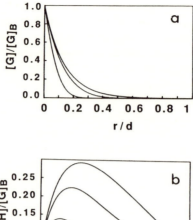

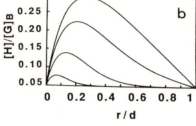

Figure 4-59. Concentration profiles for (a) glucose and (b) hydrogen peroxide in an amperometric enzyme electrode. The concentrations are normalized with respect to the bulk concentration of glucose and the position in the enzyme gel layer is normalized with respect to the total thickness $d = 94$ μm. The solution is on the left, the electrode on the right. Profiles are given (from bottom up) for 0.25, 1, 3, and ∞ s. (After Ref. 48.)

The assumption that the reaction rate is not limited by oxygen is valid only for low concentrations of glucose. It has been shown experimentally that the current is independent of the stirring rate, and this means that the current limiting mass-transport resistance is indeed located inside the gel layer.

A similar approach has been used to solve more complicated cases of two enzymes in one layer[49] and the multilayer/multienzyme model.[50] The major practical limitation arises from the self-inhibiting action of the hydrogen peroxide itself which attacks and gradually destroys the glucose oxidase. This problem is mitigated in the other two types of glucose electrode

b. Direct Oxidation of Glucose Oxidase. The natural electron acceptor for the reduced form of glucose oxidase, and for any other oxidase, is oxygen. The reduction of oxygen naturally leads to the formation of hydrogen peroxide (see Scheme 4-1, Section 4.1.4), which has an adverse effect on the stability of the enzyme. For this reason, in natural systems any oxidase is always accompanied by hydrogen peroxidase (catalase) whose primary purpose is to remove the excess of hydrogen peroxide. Catalase must obviously be removed from the glucose oxidase preparation used in the hydrogen-peroxide-based sensor described above.

Alternative electron acceptors have been used due to problems caused by the hydrogen peroxide and due to the undesirable oxygen dependence of the glucose electrodes based on oxygen sensing. It is also important to note that oxidases are one of the largest group of enzymes, and therefore improved sensors for substrates other than glucose could be developed according to this scheme. The idea is to find some way by which the reduced form of glucose oxidase $(GO)_{red}$ could be reoxidized anaerobically with the ultimate sink for the electrons being the electrode itself.

One possibility is to use low molecular weight *mediators*, which can shuttle electrons between $(GO)_{red}$ and the electrode thus bypassing the reaction with oxygen. The obvious prerequisite for this scheme is mobility of the mediator M_{ox}, which must penetrate into the interior of the enzyme, extract the electrons from the two flavin redox centers located deep inside the GO molecule (Figure 4-60), and transport them to the electrode. In the enzyme

$$FADH_2 + 2M_{ox} = FAD + 2H^+ + M_{red}$$

and at the electrode

$$M_{red} = M_{ox} + 2e$$

Various fast redox couples such as ferrocene, ferro/ferri cyanide, and ruthenium hexamine have been used for this purpose. Their redox potential must always be oxidizing (more positive) with respect to the $FADH_2/FAD$ redox couple ($E^0 = 0.05$ V at pH $= 7$). The requirement of mobility is, however, in conflict with the lifetime of the sensor. Since the mediator

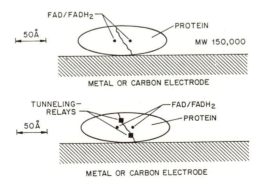

Figure 4-60. Engineered glucose oxidase with electron tunneling relays inserted. (Reprinted from Ref. 52 with permission of the American Chemical Society.)

possesses a size comparable to the substrate, it cannot be confined to the electrode proper by, e.g., a dialysis membrane. In fact, the only way this type of sensor can operate is in the sample, which contains a sufficient concentration of the mediator.[51] This requirement clearly diminishes the usefulness of such an approach for direct sensing and makes it more suitable for a sensor system.

Another possibility is to oxidize the reduced form of glucose oxidase directly at the electrode. It has been pointed out by Heller[52] that direct electron transfer between redox enzymes and electrodes is not possible for biological reasons. If it were, various enzymes with different standard potentials would equalize their potential on contact, thus effectively shorting out the biological redox chains. The "insulators" that prevent this from happening are the protein chains which surround the redox centers in these enzymes. The electron transfer rate is strongly dependent on the distance d between the electron donor and the electron acceptor,

$$k = \beta_1 \exp(-\beta_2 d) \qquad (4\text{-}150)$$

Typical values of β_1 are 10^{12}–10^{13} s^{-1}, while $0.4 < \beta_2 < 1.3$ Å^{-1}. Thus, the electron transfer rate drops by four orders of magnitude on increasing the distance from 8 to 17 Å. The radius of the GO is 43 Å, which means that communication between the electrode and the flavin center is not possible. Indeed, it has been found that GO is not electrochemically active on Pt, Au, carbon, and similar electrodes. On the other hand, when it is co-immobilized in polypyrrole and/or other electrodes made from conducting organic polymers, it transfers electrons directly to the electrode.[53] The co-immobilization is undertaken simply by electrooxidation of pyrrole (formation of polypyrrole) in solution containing also glucose oxidase. The tentative explanation for this phenomenon is that the chains of polypyrrole have sufficient mobility to penetrate the enzyme interior and to act as mediators.

The second type of electrode at which the GO is electroactive is made from insoluble organic redox salts of the donor/acceptor type, such as TTF/TCNQ (tetrathiafulvalinium/tetracyanoquinodimethanide).[54] The reason for the electrochemical activity is apparently the small but finite solubility of this salt, sufficient to cause the mediation effect.

TTF   TCNQ

The third approach to make GO electrochemically active is to "engineer" the enzyme by inserting into its interior redox molecules, which can transport electrons from the $FAD/FADH_2$ centers to the enzyme perimeter and eventually to the electrode.[52] Several molecules have been used as these electron relays, the most successful being ferrocene carboxylic acid and ruthenium pentamine $[Ru(NH_3)_5Cl_2]$. On average, one relay per 12 to 75 kdaltons of enzyme are used. This approach seems to be applicable also to other oxidases, such as aminoacid oxidase.

The amperometric response to glucose can therefore be obtained even in the absence of oxygen or any other demonstrably soluble mediator at these electrodes. These sensors are still in the early stages of their development. The main problems are related to the lifetime and possible toxicity of the redox couples used in the mediating role. It is important to note that electroactive glucose oxidase attached to the electrode fulfills the role of an electrocatalyst, which selectively increases the rate of the electrooxidation of glucose in preference to other substrates. In terms of the serial resistance model for the amperometric sensors discussed above (Figure 4-55), it is the charge-transfer resistance for glucose which is selectively decreased. The signal itself originates from the mass-transport resistance given by the geometry of the sensors. Thus, whichever approach proves to be ultimately more suitable for practical application, it is likely to be embodied in a microelectrode configuration mainly owing to its steady-state behavior.

The limited lifetime of the purified enzyme may be unacceptably short for a practical sensor, and some enzymes are not even available in their pure state. Mainly for these reasons amperometric enzyme electrodes employing cell cultures, tissues, and even whole organs have been prepared. Fundamentally they are no different from the enzyme electrodes using purified enzymes immobilized in a synthetic matrix. In general, the lifetime is indeed improved at the expense of a longer response time. The tissue or cell culture represents a diffusional barrier, which is rather difficult to control. The resistance of this barrier is sometimes such that the response of the sensor is lost. Thus, it has been found empirically that the maximum loading of bacterial cells should not exceed 1.5% of the total volume of the enzyme-containing layer.[55,56]

c. Oxygen-Based Enzyme Electrodes. The earliest design of the glucose electrode generally applicable to any enzymatic system involving oxygen was based on measurement of the relative oxygen deficiency at the oxygen electrode caused by the enzymatic reaction. The principle of operation of this sensor is shown in Figure 4-61. It consists of two identical oxygen electrodes, one (A) covered with an "active" glucose oxidase layer and the other (B) with an "inactivated" enzyme layer. The inactivation can be carried out either chemically, or by radiation, or thermally. In the absence of the enzymatic reaction the flux of oxygen to these electrodes,

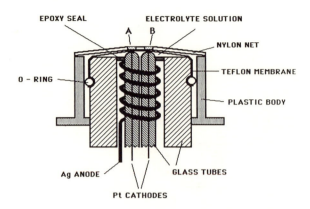

Figure 4-61. Glucose amperometric sensor based on two identical Clark-type oxygen electrodes. The gel containing active glucose oxidase is deposited at electrode A. The gel at electrode B contains only a gel or denatured (inactive) GOD.

and therefore the diffusion limiting currents, are approximately equal ($i_1 \sim i_2$). In practice, there is always some difference given by the unequal transport properties of the two layers; however, this difference can be easily corrected. When glucose is present in solution and the enzymatic reaction takes place, the amount of oxygen reaching the surface of the "active" electrode is reduced by the amount consumed by the enzymatic reaction. Therefore

$$\Delta i = (i_1 - i_E) - i_2 \tag{4-151}$$

The mathematical treatment developed for this sensor[57] is similar to that describing the hydrogen peroxide glucose electrode. There are two major problems encountered with this sensor. The first is self-inhibition caused by the hydrogen peroxide, mentioned earlier. It is less acute in this case than in the hydrogen peroxide sensor, because the catalase substantially eliminates any excess of H_2O_2. The second problem is encountered when the electrode is used *in vivo*. Under those conditions the concentration of oxygen, which is the second substrate, is low and at high glucose concentrations the current becomes limited by the availability of oxygen.

An elegant solution to this problem has been proposed (Figure 4-62).[58] By design the mass transport of oxygen takes place by cylindrical diffusion into the enzyme layer, while the transport of glucose is restricted to linear diffusion through the distal end of the sensor. Thus, a diffusional barrier (in a relative sense) is placed in the path of one substrate (glucose) while the transport of the other (oxygen) is enhanced. The steady-state mass trans-

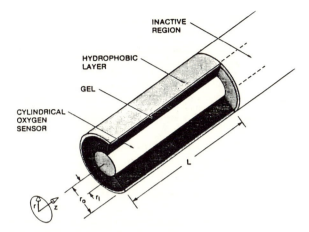

Figure 4-62. Schematic diagram of a two-dimensional cylindrical glucose electrode. (Reprinted from Ref. 58 with permission of the American Chemical Society.)

port is described by the normalized reaction-diffusion equations for glucose and oxygen in cylindrical coordinates,

$$\frac{\delta^2 C_G}{\delta\rho^2} + \frac{1}{\rho + r_i/R}\frac{\delta^2 C_G}{\delta\rho} + \xi\frac{\delta^2 C_G}{\delta\zeta^2} - \vartheta^2 V(C_G, C_O) = 0 \qquad (4\text{-}152)$$

and

$$\frac{\delta^2 C_O}{\delta\rho^2} + \frac{1}{\rho + r_i/R}\frac{\delta^2 C_O}{\delta\rho} + \xi\frac{\delta^2 C_O}{\delta\zeta^2} - \vartheta^2 V(C_G, C_O) = 0 \qquad (4\text{-}153)$$

where

$$\rho = (r - r_i)/R, \qquad R = r_O - r_i, \qquad \zeta = z/L, \qquad \xi = R/L$$

and ϑ is the Thiele modulus containing the parameters of the glucose Michaelis–Menten scheme and the diffusion and partitioning coefficients of glucose and oxygen. The boundary conditions reflect both the axial and radial geometry of the sensor. Thus the concentration of oxygen anywhere along the surface of the cylindrical oxygen sensor is zero (Figure 4-62), and so is its gradient at the proximal end ($\zeta = 1$). At the distal end ($\zeta = 0$) the oxygen gradient is given in both the radial and axial directions, while the glucose gradient is given in the axial direction only. This is due to the fact that the diffusion of glucose to the interior of the sensor is restricted by the hydrophobic cylindrical envelope (Figure 4-62). The glucose gradient at the oxygen sensor surface is zero, reflecting the fact that no glucose can be oxidized directly by the electrode. It is also zero at the proximal end of the sensor, meaning that the reaction had been completed along the axial diffusion path. The equations are solved by two-dimensional orthogonal collocation and the concentration profiles have been computed (Figure 4-63). The sensor output is expressed as the ratio of the oxygen-dependent (i_O) and glucose-dependent (i_G) currents. It is dependent on seven dimensionless parameters whose adjustment allows optimization of the performance (Figure 4-64).

The most important outcome of the analysis and experimental verification of this model is that the troublesome dependency on oxygen can be completely eliminated. The sensor therefore responds only to glucose even in the case of a three-order-of-magnitude oxygen deficieny in the bulk. Another interesting and important aspect of this design is that it tends to mitigate the adverse effect of hydrogen peroxide on the activity of glucose oxidase; as the enzyme becomes progressively damaged the reaction zone

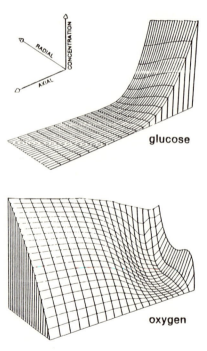

Figure 4-63. Calculated concentration profiles in a two-dimensional cylindrical glucose electrode. The sample is on the right-hand side. The surface of the electrode is at the front and the surface of the hydrophobic membrane at the back. The insulated (inactive) end of the electrode is on the left. The calculations were made for [bulk glucose] = 0.50 mM and [bulk oxygen] = 0.11 mM. (Reprinted from Ref. 58 with permission of the American Chemical Society.)

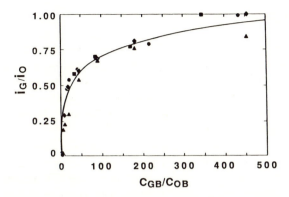

Figure 4-64. Normalized sensor current plotted as a function of the ratio of the bulk concentrations of glucose and oxygen. The different points correspond to four different concentrations of bulk glucose from 0.01 to 0.11 mM. (Reprinted from Ref. 58 with permission of the American Chemical Society.)

moves deeper into the interior of the sensor. As expected the time constant is affected, but not in a significant way.

The three types of glucose electrode discussed here illustrate the major facets of the design and operation of enzymatic amperometric sensors. Amperometric enzyme electrodes for other substrates described in the recent literature are given in Table 4-9. The actual design details of these sensors depend on the enzyme kinetics involved and on the operating conditions under which they are used.

Table 4-9. Examples of Amperometric Enzyme Electrodes[a]

Substrate	Enzyme	Sensor	Stability	Response time	Range (M)
Glucose	Glucose	$Pt(H_2O_2)$	>14 mo	1 min	10^{-2}–10^{-4}
	oxidase	$Pt(O_2)$	>4 mo	1 min	10^{-1}–10^{-5}
		$Pt(I^-)$	>1 mo	2–8 min	10^{-3}–10^{-4}
L-amino	L-AA	$Pt(H_2O_2)$	4–6 mo	2 min	10^{-2}–10^{-4}
acids	oxidase	Gas (O_2)		2 min	10^{-2}–10^{-4}
		$Pt(I^-)$	>1 mo	1–3 min	10^{-2}–10^{-4}
Succinic	Succinate	$Pt(O_2)$	<1 w	1 min	10^{-2}–10^{-4}
acid	dehydrogenase				
Aliphatic	Alcohol	$Pt(O_2)$	>4 mo	30 s	10^{-2}–10^{-4}
acid	oxidase				
Alcohols	Alcohol	$Pt(O_2)$	>4 mo	30 s	10^{-2}–10^{-4}
	oxidase				
Uric acid	Uricase	$Pt(O_2)$	4 mo	30 s	10^{-2}–10^{-4}
Phosphate	Phosphatase/	$Pt(O_2)$	4 mo	30 s	10^{-2}–10^{-4}
	glucose oxidase				

[a] Selected from Ref. 27 with permission of Wiley.

4.2.1.2. Modified Electrodes

Much research has been conducted into the chemical modification of electrodes and this subject has been reviewed extensively.[59] The motivation for the work has been to introduce additional flexibility in the design of and additional control over the electrochemical processes taking place at the electrodes. In order to achieve this goal traditional electrode materials have been replaced by "electrode superstructures" designed to facilitate a specific task. Thus, various catalysts have been attached to the electrode to lower the overall activation energy barrier of the electron transfer for specific species and/or promote long-term chemical stability of the electrode material.[60] This aspect is, of course, common with the issue of chemical selectivity in amperometric sensors. However, because the primary goal of chemical sensing is to obtain information about the chemical composition, only some of the accomplishments of modified-electrode research are relevant. This is particularly true when the main purpose of electrode modification has been to achieve greater energy efficiency of the electrochemical process, for example, in electrochemical solar-energy conversion. The amperometric biosensors discussed in the previous chapter can be viewed as a special case of superstructure electrodes.

It must be remembered that with amperometric sensors the analytical information is obtained from the mass-transport limiting current. One important consequence of the current–voltage equation [see equation (4-4) and Figure 4-3] is that one can always apply a sufficiently high potential in order to transfer electrons to/from the electrode to a given species. The problems is that, at a high electrode potential, the other species present in the solution begin to interfere; in other words, for a higher applied potential the selectivity of the sensor is diminished.

In principle there are three ways in which the chemical selectivity of the amperometric sensor can be obtained: by modification of the charge-transfer kinetics, by selective manipulation of the transport properties, and by selective accumulation of the determinand.

Selective facilitation of the *charge transfer* pertaining to the species of interest is called *electrocatalysis.* In such a case the species of interest will be oxidized/reduced at a potential substantially lower than that of the interferants. The higher selectivity therefore implies a lower applied potential at the modified working electrode, which exhibits selective electrocatalytic properties. In such a situation the choice of experimental conditions, including the type of modifying layer, is dictated by the presence of specific interferants. The bioamperometric enzyme electrodes discussed in the previous subsection belong to this category.

One can hinder selectively the access of a species or a group of species to the electrode surface. This principle is illustrated by the example of a

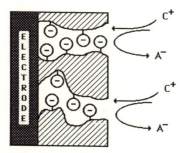

Figure 4-65. Electrostatic rejection of anions by an amperometric electrode modified with Nafion (polysulfonated polyfluorocarbon).

Pt microelectrode coated with a thin film of Nafion (Figure 4-65). Nafion is a fluorosulphonated Teflon and is highly permeable to cations and neutral species. Such electrodes have been used successfully for *in vivo* monitoring[60] of cationic neurotransmitters (such as dopamine, norepine-phrine, and 5-hydroxytryptamine) in the presence of anionic interferants, ascorbate, 3,4-dihydroxyphenylacetic acid, and 5-hydroxyindoleic acid, all of which are negatively charged at physiological pH. These anions are excluded *electrostatically* from the electrode surface by the negatively charged sulfonic groups present in the pores of Nafion. It has been shown that transport of the interfering anions to the electrode can be inhibited to 0.5–0.7% of the equivalent concentration of dopamine (Figure 4-66).

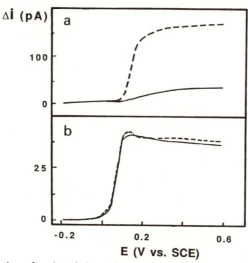

Figure 4-66. Rejection of an interfering anion by the Nafion membrane deposited over the microelectrode. The untreated electrode (a) was used in a differential pulse mode; the concentration of dopamine was 50 μM. The Nafion-treated electrode (b) was used in a normal pulse mode and $[DA] = 10$ μM. (The figure was kindly provided by M. Wightman.)

The possibility of selective, reversible binding of the species of interest to the modified layer (Figure 4-67) has been demonstrated with poly(vinylferrocene)/4-vinyl-4′-methyl-2,2′-bipyridine copolymer.[62] 2,2′-bipyridine is a complexing ligand which selectively and reversibly extracts Fe^{2+} ions from the solution. This process is in effect selective sorption by which the species of interest is accumulated at the electrode and then transformed electrochemically. The quantitative relationship between the bulk activity and concentration at the electrode interface is governed by extraction equilibrium in the case of bulk–bulk interaction, or by the adsorption isotherm if the layer is thin and the determinand is only adsorbed at the surface. In either case the surface immobilized complex of Fe–(bipyridine)₃ is then determined electrochemically. For this we have a choice of several electrochemical techniques, all derived from the basic current–voltage–concentration relationship (Section 4.1). The actual implementation of the excitation voltage (or current) may further enhance the sensitivity and selectivity of the measurement. A detailed discussion of these techniques can be found in specialized textbooks.[4] In the original paper the authors chose square-wave voltammetry and differential-pulse voltammetry, both of which are extremely sensitive electroanalytical techniques. In both cases, as the potential of the electrode is scanned a peak current is obtained (Figure 4-68) that is directly proportional to the concentration of Fe^{2+}. If two or more electroactive species are present at the electrode surface, then the corresponding number of peaks is obtained. Moreover, if the standard potentials for the oxidation (reduction) of those species were sufficiently separated ($\Delta E^O > 100$ mV), then scanning of the potential would provide additional selectivity. Thus, the overall selectivity of this sensor is due to the combination of two factors: selective complexation of Fe^{2+}/ with 2,2′-bipyridine and oxidation of the complex at the characteristic potential ($+1.0$ V vs. SCE).

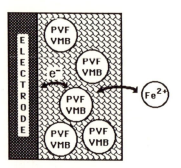

Figure 4-67. Oxidation of Fe^{2+} mediated selectively by complexation with poly(vinyl-ferrocene)/4-vinyl-4′-methyl-2,2′-bipyridine copolymer deposited at the surface of an amperometric electrode.

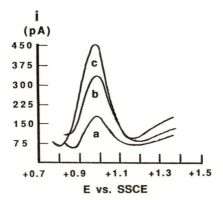

Figure 4-68. Square-wave voltammograms for a platinum microelectrode modified with poly(vinylferrocene)/4-vinyl-4'-methyl-2,2'-bipyridine copolymer. The electrode was immersed in Fe(II) solution of (a) 4.75×10^{-8} M, (b) 4.75×10^{-7} M, and (c) 4.75×10^{-6} M concentration. (Reprinted from Ref. 62 with permission of the American Chemical Society.)

Perhaps the most popular electroanalytical technique at solid electrodes is cyclic voltammetry. The applied potential is cycled linearly between two potentials, one below and one above the standard potential of the species of interest (Figure 4-69). Therefore, in one half of the cycle the oxidized form of the species is reduced, while in the other half it is reoxidized to its original form. The resulting current–voltage relationship (cyclic voltammogram) has a shape which depends on the kinetics of the electrochemical process, on the coupled chemical reactions, and on the electrode geometry. Figure 4-69 corresponds to a reversible reduction of the soluble redox couple taking place at a macroelectrode. The peak current is directly proportional to the concentration of the electroactive species, to the area of the electrode, and to the square root of the sweep rate.

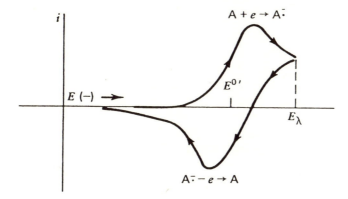

Figure 4-69. A typical cyclic voltammogram of a hypothetical redox couple A/A^-

If the electroactive species are confined to the electrode and the couple is perfectly reversible, then the peaks in the cyclic voltammogram occur at the same potential and the areas (charge) below the cathodic and anodic branches are equal. Any deviations from these conditions would clearly be reflected in the shape of the curve. For this reason cyclic voltammetry and other electroanalytical techniques contain information not only about the concentration, but also about the chemical reactions accompanying charge transfer. If the rate of the complexation reaction and the rate of diffusion to the modifying layer is fast, the concentration can be obtained from the value of the peak current. The linear calibration between 5×10^{-8} and 5×10^{-6} M concentration has been obtained in the determination of Fe^{2+} using the above procedure.

All possibilities for modulation of the electrode potential in amperometric sensors have not yet been realized. A hint of things to come can be found in Hurrell and Abruna's paper[62] in which the determination of Ca^{2+} is also described. Interference from Mg^{2+} in the determination of low concentrations of Ca^{2+} is one problem for which ion-selective electrodes do not yet provide a satisfactory answer. Yet, it is a problem of great biological importance because both these ions play a critical role in many biological processes. There are complexing agents which have very high affinity ($> 1000 \times$) toward Ca^{2+} in preference to Mg^{2+}. One such ligand is antipyrilazo III, which is electroactive in its free form. In the original paper this ligand is ion-exchanged into the modifying polymer and the electrode allowed to equilibrate with the aqueous solution containing Ca^{2+}. After transferring the electrode to acetonitrile the electrooxidation of noncomplexed ligand is measured. Using this procedure 10^{-6} M Ca^{2+} in the presence of 10^{-3} M Mg^{2+} could be determined. The procedure in its present form requires too many steps to classify it as a sensing system rather than a sensor. However, in principle, most if not all of these steps could be eliminated, thus making it a true sensor.

4.2.2. Gas Electrodes

There is no conceptual difference between amperometric sensors for solution or for gaseous species. In both cases the electrical circuit must be closed and analytical information obtained from the known concentration dependence of the mass-transport limited current. In practice, the entire gas sensor is a self-contained unit which can operate either in gas or in condensed phase. For gas-phase application the gaseous species of interest can enter the electrolyte in the sensor usually through a semipermeable membrane. In principle some selectivity can be obtained at the gas/electrolyte barrier, but it is usually not too large. Proper selection of the operating potential, the electrolyte, and modification of the working electrode usually supplement the barrier selectivety.

In this section we take a look at three types of amperometric gas sensor: Clark-type electrodes, fuel cells, and limiting-current-type gas sensors.

4.2.2.1. Clark Electrode

The need to sense oxygen is found in many fields, from biology and medicine to industrial production and safety. Likewise, the conditions under which oxygen sensors are used range from high temperature (400–1600 °C) in gas phase to normal temperatures in liquid phase. There is an extensive literature devoted specifically to oxygen sensing.[63–65] High-temperature potentiometric sensors (Section 4.1.5.3) and high-temperature amperometric sensors (Section 4.2.2.3 below) are covered separately. In this chapter we focus on a bipolar amperometric sensor, the Clark electrode. Although the principles of operation, the design rules, the problems and their solutions apply more or less to any gas which can be reduced or oxidized (such as H_2S, NO, NO_2Cl_2, CO), the discussion here will be limited to oxygen.

The electrochemical reaction in question can be expressed as

$$O_2 + 2H_2O + 2e \Rightarrow [H_2O_2] + 2OH^-$$
$$[H_2O_2] + 2e \Rightarrow 2OH^-$$

The separate polarographic wave corresponding to the reduction of hydrogen peroxide can be observed only on a mercury cathode, where the two waves are obtained for the reduction of oxygen. On other electrodes (such as Pt, Au, or C) only one four-electron wave is obtained and corresponds to the overall reduction to $4OH^-$. The potential of the working electrode (diffusion limiting-current plateau) E usually lies between -600 and -900 mV vs. Ag/AgCl.

For a bare electrode the mass-transport conditions described in Section 4.2.1 apply. In that sense oxygen is no different than any other electroactive species. Indeed, the benefits of the small electrode size have been described by Saito,[66] who found that bare Pt ring electrodes of band thickness between 100 and 500 Å were completely insensitive to flow. Until 1956 amperometric oxygen electrodes, used mainly in biomedical applications, were bare noble-metal wires. They are called *monopolar* electrodes because an external anode is required for their operation. Needless to say they can be used only in a conducting medium and they suffer from two principal problems: flow sensitivity, and fouling-up of the electrode surface with deposits which alter the mass-transport properties and change the response characteristics. It has been found that the latter problem can be minimized or eliminated by coating the electrode with hydrophilic polymers, which allow the transport of ions but restrict contact of higher molecular weight substances (e.g., proteins) with the electrode surface. The thin hydrophilic polymers are usually not strong mechanically and may therefore cause problems in applications where the electrode must be introduced to (e.g.) tissue. The flow sensitivity can be eliminated by decreasing the size of the electrode to microns.

In the Clark electrode the electrical circuit is completed within the sensor itself by placing both the cathode and anode behind the membrane, which is permeable to oxygen (Figure 4-70). Several different materials have been used (Teflon, polyethylene, silicon rubber, etc.) Clearly, one can even obtain some degree of selectivity by choosing this membrane in line with the conditions of the application. Diffusion through this structure is more complicated. For spherical geometry the steady-state current is given by

$$i = \frac{4\pi F r_0 D_S k_S P(r_1)}{D_S(r_1 - r_0)/D_m r_1 + k_m r_0/k_S r_1} \tag{4-154}$$

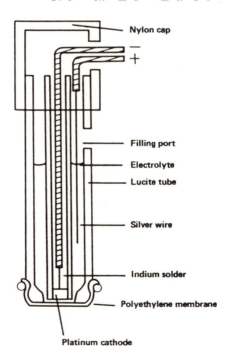

Figure 4-70. Schematic diagram of a Clark oxygen electrode. (Reprinted from Ref. 64 with permission of CRC Press.)

where r_0 is the outer radius of the membrane, D the diffusion coefficient, r_1 the radius of the electrode, k the solubility (Henry's law constant: $C = kP$) of oxygen, and $P(r_1)$ is the oxygen partial pressure at the membrane surface. Subscript S refers to the solution and m to the membrane. The concentration profile in a radial direction through the 25-μm-diameter electrode is compared in Figure 4-71 with the profile for a bare electrode. It is seen that the depletion effect caused by the electrode itself is almost completely confined within the membrane. This is particularly important if the electrode is used in the medium which itself consumes (produces) oxygen. It has been pointed out that the mutual effect of the electrode on the tissue, which itself consumes oxygen, and *vice versa*, is complicated.[64] Generally, for electrodes in which the depletion field extends beyond the

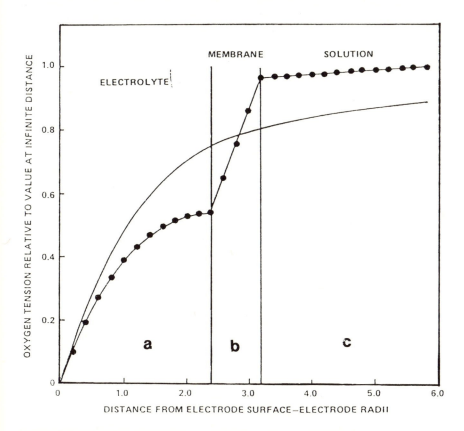

Figure 4-71. Calculated concentration profiles for oxygen in a Clark electrode consisting of (dotted curve) an internal electrolyte solution (a) and a membrane (b). The profiles extend to the sample solution (c). The smooth curve is for a bare cathode of the same dimensions. (Reprinted from Ref. 64 with permission of CRC Press.)

membrane, it is not possible to use calibration in aqueous solutions for measurements in oxygen-consuming tissue. The effect of the depletion field disappears for microelectrodes of 10-μm radius or less.

The flow sensitivity of the electrode has the same origin, as noted previously (Section 4.2.1). A stagnant (Prandtl) boundary layer of thickness δ forms around the spherical electrode (radius L) placed in the liquid (kinematic viscosity v) moving with velocity U,

$$\delta = (vL/U)^{1/2} \tag{4-155}$$

If the depletion layer lies completely inside the stagnant layer, the current will be unaffected by the change of flow. From equation (4-141) we know

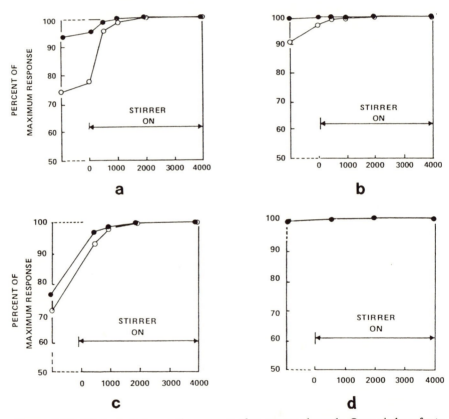

Figure 4-72. Effect of stirring on the response of an oxygen electrode. Open circles refer to O_2-saturated water and full circles to areated water. A 50-μm-diameter Pt cathode was used with (a) 6.25-μm-thick and (b) 25-μm-thick Teflon membranes. Panel (c) refers to a 300-μm-diameter cathode and panel (d) to a 10-μm-diameter cathode, both covered with a 6.25-μm Teflon membrane. (Reprinted from Ref. 64 with permission of CRC Press.)

that this will happen when the electrode radius becomes small. For a Clark-type electrode the flow insensitivity is obtained for even larger diameters of the electrode owing to the confining effect of the membrane, which has lower oxygen transmissivity $D_m k_m$ than the solution,

$$D_m k_m < D_s k_s \qquad (4\text{-}156)$$

This effect is demonstrated in Figure 4-72, where the effect of liquid velocity is compared for several sensors of different membrane dimensions and electrode radii.

It should be expected that the response speed again scales with the electrode radius. However, it has been found[67] that the fastest response speed for a semicylindrical Clark electrode is obtained for a radius of approximately 5–10 μm. This is because, as the radius decreases, the effect of the layers closer to the electrode surface becomes relatively more important than those farther away. In fact, with further decrease in a time response slower than that for planar electrodes can be obtained.

The greatest impact of amperometric oxygen electrodes has been felt in medicine and physiology. A schematic diagram of a catheter-size Clark electrode is shown in Figure 4-73. A diagram of a temperature- and pressure-compensated Clark electrode for oceanographic measurements up to 600 ft is shown in Figure 4-74. The normal temperature coefficient of the Clark electrode is about 2% per °C and the linearity is usually better than

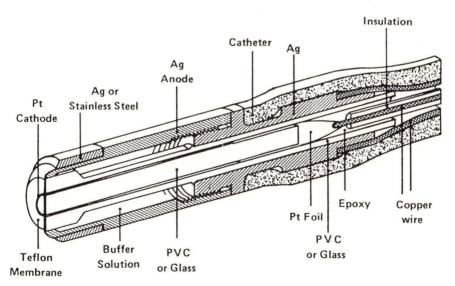

Figure 4-73. Catheter oxygen electrode with external diameter 2 mm. (Reprinted from Ref. 64 with permission of CRC Press.)

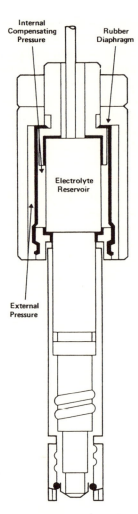

Figure 4-74. Diagram of a temperature- and pressure-compensated oxygen electrode designed for operation under high-pressure (200 m water) conditions. (Reprinted from Ref. 64 with permission of CRC Press.)

1%. The time response depends mainly on the thickness of the membrane; for a 5-μm-thick polypropylene membrane on a 7-μm-radius hemispherical electrode, the response time is below 1 s.[62]

4.2.2.2. Fuel Cells

If the number and type of gaseous species to be detected is defined narrowly and is invariable, one can sometimes use a brute-force approach to construct a chemical sensor without detailed knowledge of the mechanism of its operation. In such a case the output of the sensor, be it voltage, current, or any other physical parameter, is optimized empirically *by design* to yield a calibration curve within the expected range of concentration of the species of interest. The main prerequisite is that the calibration be stable. Serendipity, rather than adherence to the laws of physics and chemistry, plays the most critical role in their design, so it is difficult to explain the principles of operation of these sensors.

If an electrochemical cell is constructed and contains a solvent/supporting electrolyte combination with as broad a window of potentials as possible (Figure 4-1), a voltage can be applied at the working electrode such that one species can be oxidized (reduced) with some degree of selectivity. The electrochemical process of interest can be further enhanced by electrocatalysis (cf. Section 4.2.1). Therefore, the parameters which can be controlled and employed to optimize sensor performance are the solvent, electrolyte, additives, electrode material, and applied voltage. An additional and sometimes most significant improvement in performance is obtained by evaluating the output of the whole array of these sensors constructed with slightly different parameters. Again, the most important requirement is stability of their response. An advantage of these sensors is their simplicity of construction. However, because the processes which affect the response are not fully known, it is difficult to correct problems when they arise or to make rational design changes.

The relationship between the cell current and the concentration of chemical species of interest present in a more or less constant matrix can be established empirically and optimized. The sample (usually an oxidizable species) enters the cell through porous electrodes similar to those used in fuel cells, hence the device is called a *fuel-cell sensor* although its purpose is not the conversion of chemical to electrical energy.

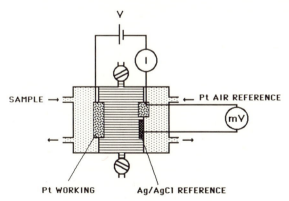

Figure 4-75. Fuel-cell-type amperometric sensor. (Reprinted from Ref. 67.)

An example is a sensor for methane[66] which also shows sensitivity to several other species. It uses the cell current as output, which is calibrated against the concentration of methane in air of 75% relative humidity. The key element in this sensor is the electrochemical cell filled with γ-butyrolactone (Figure 4-75). Examples of other solvents which have been evaluated include propylene carbonate and sulfolan. All these solvents have high boiling point (> 200 °C) and high anodic limit at the Pt electrode. Sodium or lithium perchlorate is used as the supporting electrolyte.

Propylene
carbonate   Sulfolan γ-Butyrolactone

Porous platinum/Teflon electrodes separate the electrolytic cell from the gaseous reference chamber on one side, and the sample chamber on the working-electrode side. The applied voltage between the working and Pt/air electrodes is maintained constant and the potential of the Pt/air reference electrode is measured against the potential of a regular Ag/AgCl electrode. This is a rather unusual arrangement from the electrochemical point of view because the potential of the working electrode is affected both by the cathodic processes taking place at the Pt/air reference electrode and by the IR drop across the cell. The sample enters into the electrolytic cell through the porous electrodes the pore size of which needs to be also closely controlled in order to prevent their flooding with the solvent. The electrochemical reaction of interest is postulated to be

$$CH_4 + 2H_2O \Rightarrow CO_2 + 8H^+ + 8e$$

which could explain the effect of relative humidity on the response of the sensor. Using an 800-mV applied voltage the sensor response has been evaluated in the range of 12–100% CH_4 in air. Major interference has been caused by the presence of nitrous oxide, ethane, hydrogen, and carbon monoxide. A word of caution must be added: the use of perchlorates in organic solvents always represents an explosion hazard, particularly when the assembly is, inadvertently, allowed to dry up or be exposed to higher temperatures. Other sensors of this type have been described, but they differ only in the details of their design and not significantly in concept.

4.2.2.3. High-Temperature Limiting-Current Sensors

The use of yttrium-stabilized zirconia ($ZrO_2-Y_2O_3$) as an electrolyte for the reduction of molecular oxygen at elevated temperatures (400–800 °C) has already been discussed in Section 4.1.5.3. In fact, both the reduction of oxygen and the oxidation of the oxide ion at the Pt/zirconia interface are reversible while the transport of both species in zirconia is so rapid that it is possible to construct an electrochemical oxygen pump which is the heart of the limiting-current oxygen sensor described in this chapter.[69] The overall electrochemical reaction taking place at the porous Pt electrode is

$$O_2 + 4e^- + 2V_O^+ \Leftrightarrow 2O^-$$

where V_O^+ is the oxygen vacancy in the ZrO_2 lattice. This layer is sandwiched between two porous (1-μm-thick) Pt electrodes at the bottom of a hermetically sealed cavity (Figure 4-76). As oxygen is reduced at the cathode, oxide ions diffuse through the 500-μm-thick zirconia to the anode where they are oxidized to oxygen, which exists to the atmosphere through the porous anode. Thus, molecular oxygen is pumped from one side of the sandwich to the other and the electric current which flows through the

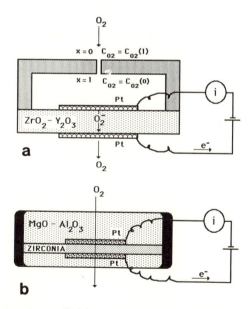

Figure 4-76. High-temperature limiting-current sensor: (a) the pinhole version and (b) the porous-layer arrangement.

circuit is limited by the oxygen flux to the cathode. In principle, it would be possible to utilize the diffusion in the zirconia itself as the mass-transport limiting process, in which case the diffusion limiting current would be directly proportional to the partial pressure of oxygen. However, it is known that when large current densities are applied the zirconia deteriorates, and the output of the sensor becomes unstable. In order to overcome this problem it is possible to place a mass-transport barrier in series with the oxygen pump. In practical sensors this is done in two ways: either a small-diameter hole is drilled in the cover (Figure 4-76a) or the pump is covered with a porous layer of, for example, MgO–Al_2O_3 (Figure 4-76b). Hence the term "limiting-current sensor" refers here to the diffusion barrier in series with the oxygen pump, and not to the diffusion limiting process in the normal electrochemical sense.

For a hole of diameter larger than the mean free path, the flux of oxygen is given as

$$J_O = -D_O A(dC_O/dx) + JC_O/C \qquad (4\text{-}157)$$

where D_O is the molecular diffusion coefficient of oxygen in the gas phase, C the total gas concentration, C_O the bulk oxygen concentration, and A is the area of the hole. The transport is considered to be unidirectional with $x = 0$ being at the orifice of the hole. The length of the hole is L. The hole is the only opening in the cavity, so the electrochemical pumping action causes a gas flow $J = J_O$. Equation (4-157) can be solved for the following boundary conditions: at $x = 0$, $C_O(x = 0) = C_O$; at $x = L$, $C_O(x = L) = C_O$. Subject to these boundary conditions, and the proportionality between the oxygen flux at the electrode and the current [see equation (4-122)],

$$i = 4FJ_O \qquad (4\text{-}158)$$

Equation (4-157) yields an expression for the steady-state current:

$$i = \frac{4FD_O AC}{L} \ln \frac{1 - C_O(L)/C}{1 - C_O(0)/C} \qquad (4\text{-}159)$$

If the oxygen concentration at the cathode $C_O(L)$ is low, then equation (4-159) becomes

$$i_1 = -\frac{4FD_O AC}{L} \ln [1 - C_O(0)/C] \qquad (4\text{-}160)$$

This corresponds to two physical situations: either the partial pressure of oxygen in the ambient $C_O(0)$ is low, in which case quantity $C_O(L)$ is also low and the current in equation (4-159) is zero, or the applied voltage is so high that every oxygen molecule reaching the cathode is reduced. This is, of course, the classical electrochemical diffusion limiting-current condition. On applying the approximation $\ln(1 - x) = -x$ for small values of $C_O(0)/C$, equation (4-160) reduces to

$$i_1 = 4FD_O AC_O/L \tag{4-161}$$

Thus for small values of oxygen concentration in the ambient the current is directly proportional to the oxygen concentration. The current–voltage characteristics of the sensor can be deduced from the following considera-tion: since the potentials at each Pt electrode are reversible, their difference can be expressed by the Nernst equation for the concentration of oxygen at the anode, $C_O(0)$, and at the cathode, $C_O(L)$. The current flowing through the layer generates a voltage drop, which is proportional to the bulk resistance of the layer, R_b:

$$V = iR_b + \frac{RT}{4F} \ln \frac{C_O(0)}{C_O(L)} \tag{4-162}$$

By subtracting equation (4-159) from equation (4-160) and again using the $\ln(1 - x) = -x$ approximation for small values of the argument, we obtain

$$i_1 - i = 4FD_O AC_O(L)/L \tag{4-163}$$

Division of equation (4-161) by equation (4-163) yields the ratio of the cathode and anode concentrations which, when substituted into equa-tion (4-162), gives the current–voltage equation for the sensor,

$$V = iR_b + \frac{RT}{4F} \ln \frac{i_1}{i_1 - i} \tag{4-164}$$

For small values of i the sensor operates in the ohmic region. For current values approaching the limiting current the applied voltage becomes infinitely high. Of course, in practice, at a high applied voltage the reduc-tion of other interfering species (such as water and CO_2) takes place. Excellent agreement between the calculated and experimental values has been observed for a wide range of temperatures (Figure 4-77).

Operation of the sensor with a porous diffusional barrier (Figure 4-76b) is similar, except that the constants have a slightly different physical meaning. Thus, S is the area of the cathode and d the thickness of the porous layer. For low concentration of oxygen

$$i_1 = 4FD_{O,eff} SC_O(d)/d \tag{4-165}$$

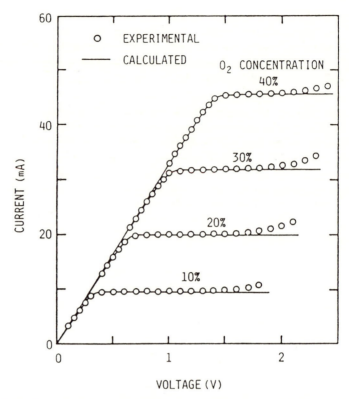

Figure 4-77. Current–voltage relationship for the limiting-current sensor. Circles are experimental points. (Reprinted from Ref. 69 with permission of The Electrochemical Society.)

The effective diffusion coefficient $D_{O,eff}$ characterizes the transport through porous medium and includes both regular diffusion and Knudsen diffusion, which has a different temperature dependence:

$$1/D_{O,eff} = 1/D_O + 1/D_{O,K} \qquad (4\text{-}166)$$

The relative contribution of these two processes is controlled by the average pore size. It is therefore possible to optimize temperature sensitivity by judicious choice of the porosity of the diffusion barrier layer.[69] The temperature coefficient of the sensor with 0.1–0.2 μm average pore diameter is 0.04% per °C between 640 and 800 °C.

The background current of the sensor is given by the contribution to the current of the electronic conductivity through the zirconia layer. It can be expressed as the equivalent oxygen concentration causing the same current. At 700 °C it is approximately equivalent to 10^{-40} atm oxygen. The contribution of the flow through the pinhole to the mass transport does not exist in the porous-layer sensor, and neither does the time constant associated with the cavity on the cathode side. For all these reasons the porous-layer version of the high-temperature limiting-current sensor has better performance characteristics than the pinhole version.

By far the most important practical use of this sensor is for automotive applications, namely, the control of the air/fuel ratio. It compares favorably with the surface-conductivity or high-temperature potentiometric sensor.[70] Other gases could be detected by employing the same principle, provided that the correct materials were used for the electrochemical pump. The electrode materials/solid electrolytes used for the construction of potentio-metric high-temperature sensors (Table 4-8) could serve as a guide.

4.3. Conductimetric Sensors

The third basic electrochemical parameter which can yield sensory information is the *conductance* $G(\omega)$ of the cell (Figure 4-1). It is related to the current and potential through the generalized form of Ohm's law [equation (D-1)]

$$G(\omega) = I(\omega)/E(\omega)$$

The unit of conductance is the siemens (mho), which is the reciprocal of the resistance R (ohm). Resistance is related to resistivity ρ by the ratio of the length L to the cross-section A,

$$R = \rho L/A \tag{4-167}$$

On the other hand, the *conductivity* σ or specific conductance (reciprocal of ρ) is normalized with respect to area, potential gradient, and time and is expressed as the ratio of the current density J (A cm^{-2}) to the electric field E (V cm^{-1}),

$$\sigma = J/E \tag{4-168}$$

Its unit is siemens^{-1}cm^{-1}. These terms will be used throughout this section interchangeably. Determination of the *volume conductivity* of solids and liquids is important for the characterization of these materials and measurement techniques have been developed in detail.

Conductimetric sensors are usually bipolar devices in which the conductance of the cell

metal electrode(1)/selective layer/metal electrode(2)

is measured at both terminals either in a bridge arrangement or, more frequently, as a current at applied voltage. These measurements can be conducted either in a DC mode or with a periodically changing excitation signal, current or voltage, according to equation (D-1). There is a limitation on the use of these sensors and it is imposed by the conductivity of the sample itself; conducting samples would obviously short out the resistance of the sensor itself. Therefore the domain of application of conductimetric sensors lies in the nonconducting samples, such as the gas phase or nonconducting liquids. In the majority of applications the relative change in the overall conductance is related to the partial pressure of the active species P_X by some general function F,

$$\partial G/G = F(P_X) \tag{4-169}$$

The general principle governing the operation of conductimetric sensors can begin to be studied by analyzing Figure 4-1. Once again we realize that the electrochemical cell is a complex arrangement of equivalent resistances and capacitances. The primary interaction between the sample and sensor involves the selective layer and the sensory information is obtained by the selective modulation of only one of these equivalent circuit elements. We must therefore consider the contact resistance (R_c) between the electrode and the selective layer, the bulk resistance R_b and the surface resistance R_s of the selective layer, and the interface resistance R_i at the interface between the selective layer and the insulating substrate (Figure 4-78). Resistance is an intensive property, so one can localize the source of the signal at one electrode by making its contact area with the selective layer smaller than that between the selective layer and the other electrode. This is exactly the same argument as used in Section 4.2 while discussing the amperometric sensing circuit.

Contact Resistance.   In all conductimetric sensors reported to date the connections are metallic (electronic conductor). This limits the possible types of contact to: metal–electronic conductor, metal–insulator–electronic conductor, metal–semiconductor, and metal–insulator–semiconductor. The metal–metal contact is generally ohmic and the contact resistance R_c will not be subject to chemical modulation.

The current density–voltage relationship for a contact in which a thin layer of insulating oxide covers the metal, and the selective layer is electronically conducting, is expected to follow the law describing the flow of a *tunneling current* at a metal–insulator–metal junction, the Fowler–Nordheim equation,

$$J = B - (\phi + V)\exp\{-4\pi d[h(\phi + V)/2mq]^{1/2}\} \tag{4-170}$$

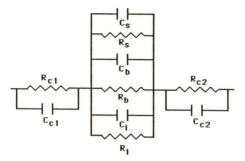

Figure 4-78. General equivalent circuit diagram of a conductimetric sensor. Subscripts c1 and c2 refer to the contacts, s to the surface, i to the interface, and b to the bulk.

where V is the applied voltage, ϕ the average barrier height, d the insulator thickness, h Planck's constant, m the mass of an electron, and B is a barrier height and thickness-dependent constant.[71] Both the barrier height and the thickness of the oxide can be modulated chemically giving rise to chemiresistive behavior.

If the selective layer is a semiconductor, a Schottky barrier may exist at the contact and may give rise to the chemically modulated resistance. The current through the Schottky barrier is a nonlinear function of the applied potential. If the metal is clean, the contact resistance is given by

$$R_c = \frac{k}{qA_R T} \exp\left(\frac{q\phi}{kT}\right) \tag{4-171}$$

where A_R is the Richardson constant for thermionic emission of electrons and given by

$$A_R = 4\pi mqk^2/h^3 \tag{4-172}$$

For a contact in which the metal is covered with its nonconducting oxide of thickness d and dielectric constant ε, the density of the Schottky current is

$$J = A_R T^2 \exp\left[\frac{-q\{\phi - (qV/4\pi\, d\varepsilon_i)^{1/2}\}}{kT}\right] \tag{4-173}$$

Such Schottky barrier junctions apparently exist in most chemiresistors which use Al as contact electrodes. On the other hand, contact electrodes made of noble metals can be prepared essentially oxide-free.

A further complication arises from the fact that, if the selective layer is an ionic conductor, the electrochemical reaction takes place at the electrode/selective layer contact. If the device is excited with a DC current, a net electrochemical reaction takes place at the contact and is governed by Faraday's law [equation (4-3)]. In such a case the mass transport of the electroactive species within the contact region must be considered. In general, when the response of the sensor depends on the chemical modulation of the contact resistance by one of the above mechanisms, it will be a strongly nonlinear function of concentration. Furthermore, because R_c always depends on the applied voltage, the optimization of the response must be carried out by examining the voltage–current characteristics of the contact.

For a time-varying excitation signal (e.g., sinusoidal voltage or current) the capacitances in the equivalent electrical circuit (Figure 4-78) must also be taken into consideration. The contact capacitance C_c, which is analogous to the double-layer capacitance of the electrolytic cell, may change as a result of the accumulation of the charged species at the contact. The bulk capacitance C_b of the selective layer may change as a result of the change in the bulk dielectric constant due to the uptake of some chemical species, such as water. The depletion layer capacitance C_s is affected by the density of the surface states and by the dielectric constant of the depletion region.

Resistance of the Selective Layer. If the contact resistance is lower than the resistance provided by the selective layer itself, the latter can be used to chemically modulate the overall resistance of the device. It can be done either through modulation of the *bulk conductivity* or through the *surface conductivity* of this layer. The resistivity of the surface layer will be discussed in connection with high-temperature oxide sensors (Section 4.3.3). The bulk conductivity σ depends on the concentration of charge carriers and on their mobility, either of which can be modulated by exposure to the gas. The first prerequisite of such an interaction is penetration to the interior of the layer. The second is the ability of the gas to enter into redox (electron-donating/accepting) equilibrium with the selective layer. Such a process then constitutes doping, which affects the overall conductivity (Figure 4-79). For a mixed semiconductor the overall conductivity is determined by the combined contribution from the holes (p) and electrons (n) as given by the general conductivity equation

$$\sigma = q\mu_p[p] + q\mu_n[n] \tag{4-174}$$

Of course, in the case of an intrinsic semiconductor for which $n_i = p = n$ equation (4-174) simplifies to

$$\sigma = qn_i(\mu_p + \mu_n) \tag{4-175}$$

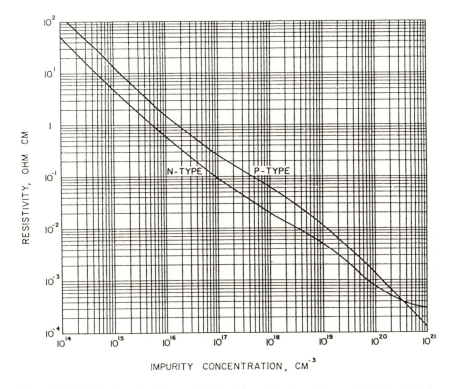

Figure 4-79. Dependence of the resistivity of p- and n-type silicon at 300 K on doping. (Reprinted from Ref. 72 with permission of Wiley.)

The modulation of charge-carrier mobility in a conducting polymer can be affected by the degree of swelling of the polymer matrix caused by vapors of some organic solvents. In some cases this process is sufficiently reversible to qualify as a possible mechanism for sensing solvent vapors.

Chemical modulation of the surface conductivity forms the basis of the commercially most successful chemical sensors, the high-temperature semi-conducting oxide sensors. The reason for their commercial success lies in the fact that their performance and cost exactly match the specific practical needs of the automotive industry. They have been described in great detail from the viewpoint of their underlying physics and chemistry.[72-74]

Generally speaking, solid-state conductimetric sensors are some of the simplest and least expensive chemical sensors known and, moreover, the supporting electronics is quite simple. The mechanism underlying the operation of some of them (such as semiconducting oxide sensors) is well understood, and their performance has been optimized. However, other conductimetric sensors are treated only phenomenologically, i.e., the relationship between the output signal and the concentration of the active species is described without a deeper understanding of its origin. The fundamental equivalence current and resistance inherent in the generalized Ohm's law [equation (C-1)] sometimes blurs the line between amperometric and conductimetric sensors. Nevertheless, it is important to remember that chemiresistors are fundamentally electrochemical devices and that both the interpretation of the results and the optimization of their design should be undertaken with the help of the general rules governing the operation of electrochemical cells.

4.3.1. Chemiresistors

One of the simplest and perhaps least discussed chemiresistors is the mercury vapor sensor. The heart of this device is a thin (75–400 Å) film of gold evaporated on a ceramic or glass substrate. At these thicknesses the films are continuous with overall resistance between 300 and 1500 Ω (sheet resistivity 2–10 Ω square). The sensors are usually connected in pairs in a DC bridge configuration with one sensor acting as reference.

When DC voltage is applied to the electrodes, the conduction electrons travel in the thin film between the two electrodes. Their mean free path is comparable with the thickness of the film. When a foreign species is adsorbed at the Au film surface, it acts as a scattering center for the electrons thus effectively increasing the resistance of the film in the surface layer (Figure 4-80). This is equivalent to impurity (or carrier) scattering observed in the bulk of doped semiconductors,[75] except that in this case the scattering centers are formed by the adsorbate. For thin films the contribution of the surface layer, which is affected by adsorption, is significant as compared to the contribution to the overall conductance from the interior.

There are several gaseous species which have high affinity for a gold surface: atomic mercury, halogens, and hydrogen sulfide. The dynamic range depends on the adsorption isotherm and the sensitivity is affected by the film thickness. For thicker films, the contribution of the above effect to the overall change in resistivity would evidently be reduced.

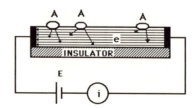

Figure 4-80. Diagram of a thin gold film chemiresistor.

Although in the original paper[76] the response is described as "reversible," the actual commercial instrument is used in an integrating mode followed by thermal desorption at 150 °C. The dynamic range for mercury is 1–1000 pg m^{-3} and for H_2S, 3–500 ppb. The response time is 13 s. There is practically no interference from water, oxygen, SO_2, or hydrocarbons.

Most other types of chemiresistor are constructed on an almost identical principle. The chemically selective layer, which can be, e.g., a thin film of organic semiconductors (0.01–1 μm), is applied over the interdigitated electrode array (Figure 4-81) situated on an insulating substrate (quartz, glass alumina) by evaporation (sublimation), as a Langmuir–Blodgett film, or by solvent spin-casting. Electrode spacing is typically 10–100 μm and the total electrode area is typically 0.1–1 cm^2. The overall resistance of these films is on the order of 10^8–10^{11} Ω. For a typical applied voltage of 1–5 V the currents are in the nA range and, more importantly, the *current density* is in that range too. This means that if the signal arises from the chemical modulation of the contact conductance, the device is usually operating in the kinetic regime (Figure 4-1). This alone would cause the response to be nonlinear with concentration. Moreover, the effect of a serial resistance of the chemically sensitive layer is to flatten the current–voltage characteristic.

A variety of electrode materials have been used ranging from transition metals (such as Ni, Al, Ti or Cu) which are covered with surface oxides, to oxide-free noble metals (such as Au, Pt, or Ag). They do not readily form oxides, so the adhesion of noble metals to the insulating substrates is generally poor. In order to overcome this problem a thin layer (500–1000 Å) of "adhesive" metal (Cr, Ti, W) is used below the noble metal. Invariably, the "adhesive" metal migrates through the noble metal and forms an oxide layer, which significantly alters the surface properties of this metal. This is a general problem affecting all planar structures that employ thin films of noble metals.[77] Schottky barrier-type behavior has been observed for chemiresistors utilizing a metal/phthalocyanine (1000 Å)/metal combination[78] that respond to NH_3. However, with both electrodes of comparable size one cannot unravel the origin of the response

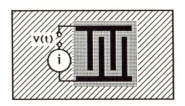

Figure 4-81. Schematic diagram of an interdigitated chemiresistor with a chemically selective layer deposited over the electrodes and with a time-varying voltage source and a current meter.

with respect to the polarity of the applied voltage; in that case either electrode/selective layer contact could be contributing in a comparable way. In this respect the sensor proposed by Pons and co-workers[79] represents a qualitative departure from the conventional chemiresistor design. It is based on two concentric-ring ultramicroelectrodes of unequal size (area ratio, approximately 10^6) (Figure 4-82). The smaller one is the working electrode and the larger one is the auxiliary (pseudoreference) electrode (Section 4.2). The pA current flows over the surface of the insulating ring separating the two electrodes. Hence the insulator surface can be viewed as the selective layer. Obviously it can be modified by the deposition of additional layers. When the device is exposed to electroactive gas, reduction/oxidation takes place at the working electrode. Therefore the origin of the signal is the contact resistance of the working electrode. Fast response in the ppm range to a variety of amines, esters, amides, nitro compounds, and so on has been reported.

The change in surface conductivity during the *adsorption* of NO_2 is postulated to be the source of the signal in a lead phthalocyanine chemiresistor.[80] An approximately 3000-Å-thick phthalocyanine layer has been deposited over an interdigitated grid of Pt electrodes. At 150 °C the device responded reversibly between 1 ppb and 10 ppm of NO_2. The response was independent of the film thickness, so it has been concluded that the response is due to the adsorption. However, the thickness insen-

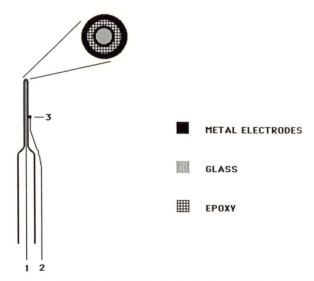

METAL ELECTRODES

GLASS

EPOXY

Figure 4-82. Amperometric ultramicroelectrode used as a gas sensor. The inner ring of 100 μm diameter and 100 μm radius is the working electrode (1) and the outer metal ring is the auxiliary electrode (2). Contact (3) is made by soldering copper wire to the surface of the outer ring. (After Ref. 77.)

sitivity could also be due to modulation of the contact resistance. This seems to be the case with chemiresistors utilizing Langmuir–Blodgett films of structurally similar porphyrins and Cu interdigitated electrodes.[81]

4.3.2. Semiconducting Oxide Sensors

The principle subject to which these sensors operate can be summarized as follows. There is a finite density of electron donors (e.g., adsorbed hydrogen) and/or electron acceptors (e.g., adsorbed oxygen) bound to the surface of a wide-bandgap semiconducting oxide, such as SnO_2 or ZnO. They represent surface states which can exchange electrons with the interior of the semiconductor, thus forming a space-charge layer situated close to the surface (Figure 4-83). The position of the surface state relative to the Fermi level of the semiconductor depends on its affinity to electrons. If its affinity is low, it will lie below the Fermi level and donate electrons (reducing agent) to the space-charge region. If it is an acceptor, it will be positioned above the Fermi level and extract electrons (oxidizing agent) from the space-charge region. By changing the surface concentration of the donors/acceptors the conductivity of the space-charge region is modulated. In an n-type semiconductor, such as SnO_2 and ZnO, the majority carriers are electrons therefore the change in surface conductivity $\Delta\sigma_s$ is given from equation (4-175) in the form

$$\Delta\sigma_s = e\mu_s \, \Delta n_s \qquad (4-176)$$

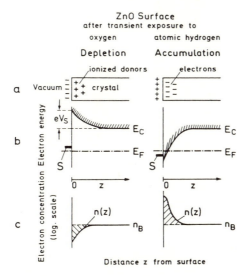

Figure 4-83. Space-charge modulation by adsorption in a semiconducting n-type oxide sensor. (Reprinted from Ref. 73 with permission of Elsevier.)

where e is the electron charge and μ_s is the electron mobility at the surface. The excess density Δn_s of charge carriers in the space-charge region of thickness d is obtained by integrating the difference between the electron density in the space-charge region and in the bulk (n_b) over the thickness d,

$$\Delta n_s = \int_0^d [n(z) - n_b]\, dz \tag{4-177}$$

The change in surface conductance ΔG_s is then

$$\Delta G_s = \Delta \sigma_s W/L \tag{4-178}$$

For an n-type semiconductor (such as ZnO) an increase in the surface concentration of the electron donor (e.g., hydrogen) will therefore increase the conductivity, and *vice versa*. On the other hand, adsorption of hydrogen on a p-type oxide (such as CoO) will decrease its conductivity by the same argument.

The bulk conductance which is not modulated by the surface reactions is represented by a parallel conductor (Figure 4-84),

$$G_b = n_b e \mu_b W d/L \tag{4-179}$$

The terms in equation (4-179) have the same meaning as those in equations (4-176) and (4-178). Subscript b refers to the bulk quantities while d is the total thickness of the oxide layer. Since the overall conductance is measured, it is advantageous to have the films as thin as possible.

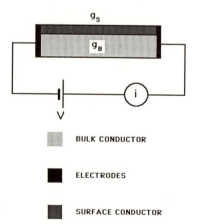

Figure 4-84. Schematic representation of a semiconducting oxide sensor with bulk (g_B) and surface (g_S) conductivities shown. (After Ref. 72.)

The relative change in the conductance of the whole device is obtained from equations (4-176), (4-178), and (4-179) and by assuming that $\mu_b \sim \mu_s$. Hence

$$\Delta G/G = \Delta n_s/n_b \, d \qquad (4\text{-}180)$$

For a high value of the relative conductance change (high sensitivity) it is necessary to have a low density of bulk carriers and a thin film. The space-charge thickness is typically 100–1000 Å, depending on the doping. As an example let us consider a typical value of the excess surface-state density $n_s = 10^{12}$ electrons cm^{-2} and the bulk density $n_b = 10^{17}$ electrons cm^{-3}. This means that for a 100-μm-thick film the sensitivity $\Delta G_s/G_b = 10^{-3}$. However, for $d = 100$ Å the sensitivity is 10 and for thin films (< 1000 Å) the space-charge region extends throughout the whole film thickness.

It is convenient to present the elemental aspects of the sensing mechanism assuming a regular crystal lattice, i.e., a monocrystal. Indeed the fundamental studies into reactions at oxide surfaces were conducted on well-defined surfaces. On the other hand, practical sensors are prepared by techniques which yield polycrystalline layers. For the dimensions of the film at which the surface conductance begins to dominate the overall conductance, the film morphology becomes the most important parameter. A schematic representation of three types of oxide surface is shown in Figure 4-85.

Although any adsorbate capable of being involved in the electron exchange with the surface can act as the surface state which dominates the space charge, oxygen and hydrogen are by far the most important species. The chemical interactions leading to changes in the density of surface states are complex and form part of the surface catalytic processes. In addition to the above-mentioned interaction of adsorbates, the diffusion of lattice defects from the bulk of the crystal also takes place. The defects can act as donors or acceptors. Oxygen vacancies V_{O^+} are presumed to be intrinsic donors,[72]

$$(\mathrm{Zn}^{2+}\mathrm{O}^{2-})_{\mathrm{lat}} = V_{O^+} + e^- + \tfrac{1}{2}\mathrm{O}_2(\mathrm{gas})$$

and

$$(\mathrm{Zn}^{2+}\mathrm{O}^{2-})_{\mathrm{lat}} = V_{O^{2+}} + 2e^- + \tfrac{1}{2}\mathrm{O}_2(\mathrm{gas})$$

while adsorbed O_2 and O species act as electron acceptors. Their relative population at the oxide surface to a large extent determines the type of chemical interaction which will take place. The issue of the selectivity of these surfaces has been treated comprehensively and in great detail elsewhere[73,74]; here only a few examples will be mentioned.

SCHOTTKY BARRIER

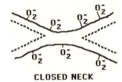

CLOSED NECK

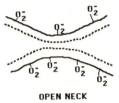

OPEN NECK

Figure 4-85. Effect of the morphology of a semiconducting oxide on conductivity. The depth of the space-charge region is indicated by the dashed line. The Schottky-barrier model assumes a tunneling mechanism for the electrons. Conductivity of the closed-neck and of the open-neck sample is modulated by the density of surface states (O_2^-).

Reactions of the crystal surface involving molecular hydrogen are important not only from the standpoint of hydrogen sensing, but also for all reactions in which the abstraction of the hydrogen atom is part of the catalytic mechanism. At low temperatures adsorbed hydrogen is the source of additional electrons in the space charge

$$H + (ZnO)_{lat} = (ZnOH^+)_{lat} + e^-$$

while at higher temperature the lattice becomes reduced

$$2H + (ZnO)_{lat} = H_2O + Zn^0$$

Another ubiquitous species, water, is also involved in the surface redox chemistry of SnO_2

$$H_2O + (SnO_2)_{lat} = (HO-Sn^{2+})_{lat} + (OH^-)_{lat} + e^-$$

The oxidation and degradation of small hydrocarbon molecules has many features in common (Figure 4-86). These reactions have been studied extensively by a variety of surface techniques, such as thermal desorption (TDS), ion mass spectroscopy, and specular infrared spectroscopy. The reaction pathway and final product depend on the type of oxygen-bearing species, which in turn depends on the doping and morphology of the oxide layer. This is the principal reason why results obtained with different oxide sensors in different laboratories do not always agree. The complexity of the surface chemistry involved can be illustrated by the example of oxidation of methane on a (101) SnO_2 surface,

$$CH_4(g) = CH_3(s) + H(s)$$

where g and s refer to gas and surface-bound species, respectively. At higher concentrations two adjacent methyl groups combine and react with the lattice oxygen,

$$2CH_3 + O_{lat} = CH_3CH_2O(s) + H(s)$$

The surface-bound ethoxide ion is dehydrated and desorbed from the surface as ethylene, leaving behind an oxygen vacancy,

$$CH_3CH_2O(s) + H(s) = H_2O(g) + C_2H_4(g) + V_O$$

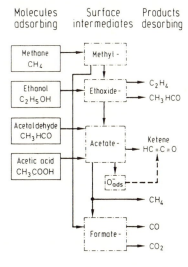

Figure 4-86. Catalytic decomposition schemes of some organic molecules relevant to semi-conducting oxide sensors. (Reprinted from Ref. 73 with permission of Elsevier.)

An alternative pathway may lead to the formation of surface-bound acetate by the second oxidation step involving lattice oxygen,

$$CH_3CH_2O(s) + O_{lat} = CH_3CH_2OO(s) + 2H(s)$$

Had the first oxidative coupling not taken place, the final product would have been adsorbed formate,

$$CH_3(s) + 2O_{lat} = HCOO(s) + 2H(s)$$

All the above species have been detected in various quantities at oxide surfaces. A discussion of this example serves mainly to show that catalytic reactions at oxide surfaces are very complex. This is a mixed blessing from the sensing point of view. On the one hand, it provides a broad spectrum of reactions which could be used. On the other hand, it can lead to great variation in the results obtained with only slightly different sensors. Another drawback of such a complex and diverse mechanism is the relatively slow time response which, in most cases, is limited by the rates of the chemical reactions (Figure 4-87). Naturally, one tendency of current

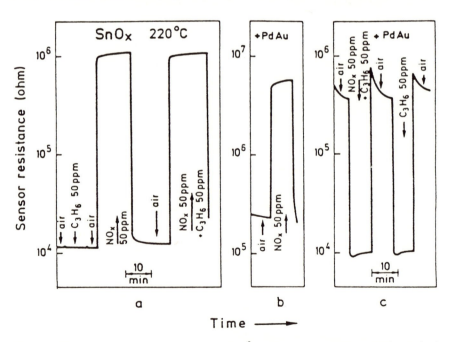

Figure 4-87. (a) Response of a thin-film (~ 1000 Å) SnO_2 sensor to NO_x and ethane. (b, c) Response to the same gases of the SnO_2 sensor with a discontinuous layer of Pd/Au (350 Å) deposited on the top. (Reprinted from Ref. 72 with permission of Elsevier.)

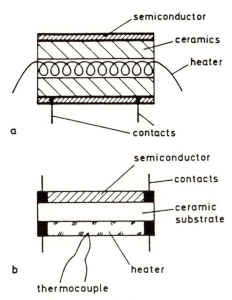

Figure 4-88. Two versions of the SnO_2 sensor built on (a) a ceramic tube and (b) a planar insulating substrate. (Reprinted from Ref. 72 with permission of Elsevier.)

research in this field is to increase the selectivity of the surface reactions by introducing additional catalytic control, e.g., by the incorporation of catalytic metals, metal clusters, and other surface modifiers.

When elevated temperature is required for its operation, an integral heater can be incorporated in the design of the sensor (Figure 4-88). One can measure either the steady-state value of the resistance or, for slower

Table 4-10. Response of a Commercial SnO_2 Sensor to Organic Vapors in Air[a]

Gas	Formula	Concentration (ppm)
Hydrogen	H_2	1,000
Methane	CH_4	20,000
Ethylene	C_2H_4	4,000
Ethane	C_2H_6	20,000
Ethanol	C_2H_5OH	10,000
Propane	C_3H_8	4,000
Butane	C_4H_{10}	20,000
Carbon dioxide	CO_2	40,000
Carbon monoxide	CO	4,400
Methyl mercaptan	CH_3SH	11
Sulfur dioxide	SO_2	10,000

[a] Reprinted from Ref. 72 with permission of Elsevier.

reactions, the initial slope as a function of concentration. The principal gases for which this type of sensor has found most widespread application are the lower hydrocarbons. However CO, SO_2, H_2, NO_x, and other gases have been measured. Table 4-10 lists the selectivity of a commercial sensor to some of these gases.

4.3.3. Conductimetric Gas-Membrane Sensors

The types of gas sensor in which the sample (gas) is separated from the internal sensor by a thin gas-permeable membrane have been discussed in previous sections. Any type of internal sensor can be used in this configuration and conductimetric sensors are no exception.[82]

The theory governing the operation of a conductimetric gas-membrane sensor has been verified in detail experimentally for CO_2 and SO_2, but sensors for H_2S and NH_3 based on the same principle have also been constructed. The main transport and equilibration processes are the same as in the Severinghaus electrode [see equation (4-88)]. Each dissociated species contributes to the overall conductivity, hence the specific conductivity κ of the cell is given by

$$\kappa = 10^{-3}(\lambda_{H^+}a_{H^+} + \lambda_{HXO_3^-}a_{HXO_3^-}) \tag{4-181}$$

where the λ quantities are the equivalent ionic conductances and HXO_3^- is the anion. The composition of the internal solution does not vary significantly, so it is legitimate to use concentrations instead of activities. By an analogous argument to that for the Severinghous electrode, the specific conductivity in the case of a cell containing initially pure water is [see equations (4-89) and (4-181)]

$$\kappa = 10^{-3}(\lambda_{H^+} + \lambda_{HXO_3^-})(KP_{XO_2})^{1/2} \tag{4-182}$$

The experimental conductance S of the cell depends on its geometry as characterized by the cell constant $\Lambda = \kappa/S$. Therefore the output of the sensor is related to the partial pressure of the gas by the expression

$$S = 10^{-3}K^{1/2}[(\lambda_{H^+} + \lambda_{HXO_3^-})/\Lambda](P_{XO_2})^{1/2} \tag{4-183}$$

The magnitude of the dissociation constant plays an important role in the response characteristics of the sensor. For a weakly dissociated gas (such as CO_2, $K_a = 4.4 \times 10^{-7}$) the sensor can attain its equilibrium value in less than 100 s and no accumulation of CO_2 takes place in the interior layer. On the other hand SO_2, which is a much stronger electrolyte ($K_a = 1.3 \times 10^{-2}$), accumulates inside the sensor and its response time is large. One way of overcoming these difficulties is to incorporate the conductivity cell within a fluid circuit (Figure 4-89) in which the water is recirculated through a mixed ion-exchanger bed whose purpose is to capture the ions and restore the conductivity baseline. In that case the measurement is carried out at a precisely timed interval after stopping the pump. This approach is really a sensor system despite the fact that it is sufficient to operate the pump manually and intermittently.

A continuous sensor has been constructed taking advantage of the principle of nonzero flux. It again uses a mixed bed of ion-exchange resin (4) (Figure 4-90a) which acts as a sink for the ions. It is separated from the gas-permeable membrane (1) with imprinted conductimetric electrodes (2) by a mesh spacer (3). The gas permeates through the membrane, dissolves, and dissociates into its ionic constituents. Since the ions are taken up by the ion exchanger, a concentration gradient develops through the spacer (Figure 4-90b). This situation is a direct analogy of the Clark electrode (Section 4.2.2.1) in which the determinand (oxygen) was continuously consumed by the cathode.

The detection limit and sensitivity of the conductimetric gas sensors depend on the value of the dissociation constant, on the solubility of the gas in the internal filling solution, and, to some extent, on the equivalent ionic conductances of the ions involved. The performance data for SO_2 and CO_2 sensors are summarized in Table 4-11.

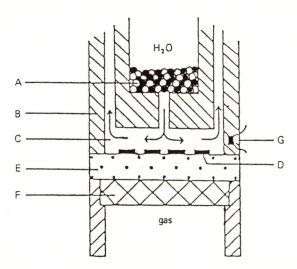

Figure 4-89. Schematic diagram of a conductimetric Severinghaus electrode with internal recirculation of the electrolyte: (A) mixed bed ion-exchanger, (B) sensor housing, (C) thin layer of de-ionized water, (D) electrodes, (E) Teflon membrane, (F) charcoal filter, (G) electrode body (Reprinted from Ref. 82 with permission of The Chemical Society.)

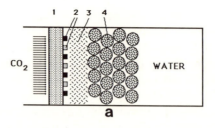

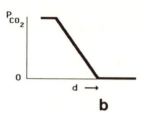

Figure 4-90. (a) Schematic diagram of a conductimetric electrode with ion-exchange regeneration. (b) The concentration gradient of CO_2. (After Ref. 80.)

Although in all conductimetric gas sensors described to data an aqueous filling solution has been used in principle, it is possible to employ any liquid for that purpose. The choice of dielectric constant and solubility would then provide additional experimental parameters, which could be optimized in order to obtain higher selectivity and/or lower detection limit.

4.3.4. Dielectrometers

Most conductimetric sensors discussed in the previous sections operate in a DC mode. This means that the capacitive equivalent circuit elements in Figure 4-78 can be ignored. When a periodically changing excitation signal is chosen for the operation of chemiresistors, it is usually in order to eliminate the contribution of the electrode polarization to the overall resistance. In contrast, dielectrometric sensors rely on the chemical

Table 4-11. Calibration Data for Carbon Dioxide and Sulfur Dioxide[a]

Species	Pressure (ppm)	Slope (s.d.) $(\Omega^{-2}(ppm)^{-1})$	Intercept (s.d.) (Ω^{-2})
SO_2	0.0–1.00	$6.39(0.337) \times 10^{-11}$	$0.11(0.078) \times 10^{-11}$
SO_2	0.00–0.20	$6.75(0.219) \times 10^{-11}$	$-0.03(0.03) \times 10^{-11}$
CO_2	0–10,000	$10.6(0.25) \times 10^{-15}$	$0.09(0.08) \times 10^{-15}$

[a] Data taken at 120 s; gas flow rate, 0.5 dm^3 min^{-1}. (Reprinted from Ref. 82 with permission of The Chemical Society.)

modulation of one or more equivalent circuit capacitors, usually through the change in the dielectric constant of the chemically sensitive layer.

Owing to its high dielectric constant (78.5 at 25 °C) water is the primary, but not only, species which can be measured by following the changes in the values of the capacitive elements in the equivalent circuit. Therefore this section could be also called "hygrometry," because within it humidity sensors are the largest represented group. However, because "DC hygrometers" (water-sensitive chemiresistors) have also been constructed and because dielectrometers for other molecules can also be made, the general term dielectrometry will be retained. Humidity sensors have been reviewed by Yamazoe and Shimizu.[83]

The simplest dielectrometer is a capacitor containing a layer of material which can more or less reversibly take up chemical vapor of a given dielectric constant. The diagram of such a sensor and its equivalent electrical circuit is shown in Figure 4-91.[84] There are several things that should be pointed out in this design. First, the upper electrode is porous either because it is very thin (100–200 Å) or because it is deposited under such conditions that it cracks. In any case the polymer beneath it comes rapidly into contact with the vapor. Obviously, it is difficult to make robust electrical connections to the top electrode. Fortunately, this is unnecessary because it forms an electrically floating plate which is common to the two capacitors: one between the Cr, Ni, Au plate (C_1) and the other between the top and the Ta plate (C_2). The corresponding leakage resistances are R_1 and R_2. The response of this sensor to water vapor is shown in Figure 4-92.

The next setup, but no different in principle, is the FET-based humidity sensor.[85] The gate capacitor comprises the usual SiO_2/Si_3N_4 combination of solid gate insulator on top of which a 1-μm-thick layer of a (unspecified) polymer is applied. This polymer is sandwiched between

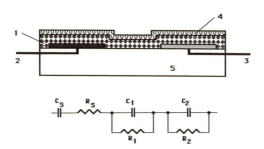

Figure 4-91. Schematic diagram and equivalent electrical circuit for a capacitive humidity sensor: (1) cellulose acetate/butyrate polymer, (2) tantalum electrode, (3) Cr, Ni, or Au electrode, (4) electrically floating porous electrode, (5) insulating substrate. (Reprinted from Ref. 84 with permission of Elsevier.)

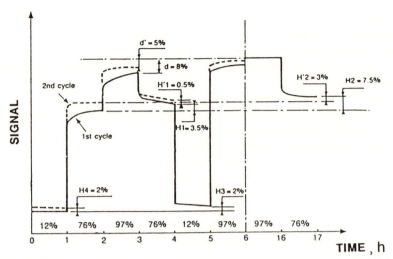

Figure 4-92. Response of a capacitive humidity sensor tested by exposure to humid atmosphere above saturated solutions of different salts. (Reprinted from Ref. 84 with permission of Elsevier.)

two gold electrodes: the top one is porous, 100–200 Å thick, and permeable to moisture. The two electrodes are connected by a shunt resistor ($R_b > 10 \text{ M}\Omega$). The equivalent circuit of this sensor is shown in Figure 4-93. A sufficiently high DC bias voltage is applied to the gate in order to turn on the FET and well into the saturation region (Section 4.1.3.2b). The input resistance of the transistor is higher than R_b, hence the applied DC bias voltage appears on the gate in addition to a small-amplitude AC voltage v_g of 10 kHz. The drain current is given by

$$I_D = \frac{\mu_n C_0 W}{2L}(V_G - V_T)^2 + \frac{\mu_n C_0 W}{L}(V_G - V_T)v_g \qquad (4\text{-}184)$$

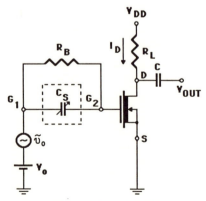

Figure 4-93. Equivalent electrical circuit for a humidity-sensitive FET. (After Ref. 85.)

The AC component of the drain current is measured as a function of the membrane capacitance C_s. A diode is incorporated on the same chip to provide temperature compensation. The accuracy of this sensor is better than 3% over the 0–100% range of relative humidity and the time response is below 30 s.

Rugged and chemically resistant humidity sensors utilizing ceramic conductors have been described. Their principle of operation is similar to polymer-based humidity sensors.[83]

The parallel-plate capacitor geometry has obvious advantages in the interpretation of results. However, they may be more than offset by difficulties with the fabrication of these devices. The deposition and mechanical stability of a 100–200-Å-thick layer of conductor on top of a polymer required to operate the above sensors is not a trivial matter.

The lateral migration of charge over the surface of a planar structure (crosstalk) is under normal circumstances a nuisance to be prevented by rigorous encapsulation, separation of the critical elements with guard rings, etc. However, it has been turned into an advantage in charge-flow transistors (CFT)[86] in which crosstalk between two transistors can be

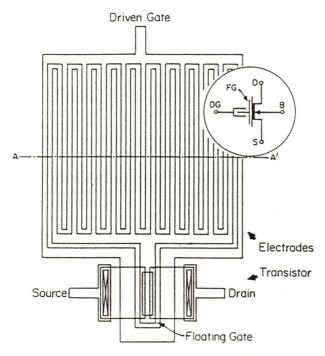

Figure 4-94. Schematic representation of a charge-flow transistor. (Reprinted from Ref. 84 with permission of the Institute of Electrical and Electronic Engineers.)

interpreted in terms of the impedance of the space separating their two gates. Two transistors, one with a driven gate and the other with a floating gate, are located on the same chip in close proximity (Figure 4-94). The material, whose permittivity changes as the result of some chemical reaction, is deposited over and between the two interdigitated gates. When the reaction takes place it modulates the fringe electric field between the driven and floating gates. The effective voltage which appears at the floating gate, V_F, as a result of the voltage applied to the driven gate, V_D, is given by

$$V_F = V_D \frac{1 + (C_X/C_T)\sinh\beta}{\cosh\beta + (C_L/C_T + C_X/C_T)\sinh\beta} \qquad (4\text{-}185)$$

where

$$\beta^2 = J(C_T/C_S)/[j + (\omega R_S C_S)^{-1}] \qquad (4\text{-}186)$$

The individual elements are explained in the equivalent circuit model (Figure 4-95). The equivalent resistance R_S and capacity C_S belong to the chemically modulated film. The phase shift and magnitude of the AC components of the two drain currents are measured between 0.1 Hz and 10 kHz. The sheet resistance R_S between 10^9 and $10^{16}\ \Omega/^0$ is hence determined with 10% accuracy.

There have been two principal applications of this device: as a humidity sensor and as an *in situ* monitor of polymerization. For humidity measurements a thin (400–800 Å) layer of Al_2O_3 is deposited over the surface of the device. The transfer function subject to equations (4-185) and (4-186) is evaluated for several frequencies (Figure 4-96). In polymerization

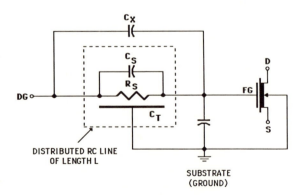

Figure 4-95. Equivalent electrical circuit of a charge-flow transistor. The voltage is provided by the driven gate (DG) and transmitted through the transmission (C_T) distributed RC line and through the sample C_X to the floating gate (FG). (Reprinted from Ref. 86 with permission of the Institute of Electrical and Electronic Engineers.)

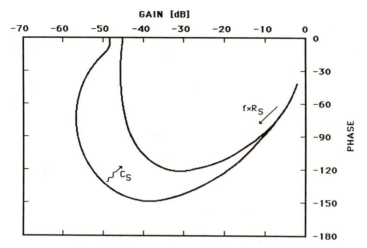

Figure 4-96. Transfer-function diagram of a CFT calculated for two different values of ratio C_T/C_S. The effect of the change in sheet resistance R_S and sample capacity C_S on the shape of the curves is indicated by arrows. (Reprinted from Ref. 86 with permission of the Institute of Electrical and Electronic Engineers.)

measurements the monomer is applied over the surface and the reaction initiated. The changes of the dielectric constant in the course of polymerization are again evaluated from the transfer-function diagrams.

Glossary for Chapter 4

A	Area
C_{dl}	Double-layer capacitance
D_i	Diffusion coefficient
E	Potential on hydrogen scale
G	Conductance
g_m	Transconductance
i	Current
i_L	Limiting current
i_0	Exchange current
I_D	Drain-to-source current
J_i	Mass flux
j_0	Exchange current density
K_{ex}	Ion-exchange equilibrium constant
k_0	Heterogeneous rate constant
K_{sp}	Solubility product
$K_{x,i}$	Potentiometric selectivity constant
L	Length
m_i	Mass-transport coefficient

P_i Partial pressure
Q Total charge
q_i Interfacial charge
r Radius
R_{ct} Charge-transfer resistance
t_i Transference number
U_i Absolute mobility
u_i Electrolytic mobility
V Applied voltage
W Width
α Solubility coefficient
χ_L Dipole (surface) potential
δ Thickness of the diffusion layer
ε Dielectric constant
ϕ Potential
γ Interfacial energy
Γ_i Surface concentration
η Overpotential
κ_D Debye length (reciprocal)
λ_i Equivalent ionic conductivity
Λ Specific cell conductance
$\tilde{\mu}$ Electrochemical potential
μ Chemical potential
μ_n Mobility of electrons in semiconductor
π Interfacial potential difference
ρ Resistivity
σ Conductivity
τ Time constant

Abbreviations Appearing in Chapter 4

CHEMFET   Chemically sensitive field-effect transistor
CWE   Coated-wire electrode
ENFET   Enzymatic field-effect transistor
FET   Field-effect transistor
FIA   Flow injection analysis
IGFET   Insulated-gate field-effect transistor
ISE   Ion-selective electrode
ISFET   Ion-sensitive field-effect transistor
MIS   Metal–insulator–semiconductor
SBT   Site-binding theory
SGFET   Suspended-gate field-effect transistor

References for Chapter 4

1. J. O'M. Bockris and A. K. N. Reddy, *Modern Electrochemistry*, Plenum Press, New York, 1970.
2. G. Kortüm, *Treatise on Electrochemistry*, Elsevier, New York, 1965.
3. J. Koryta, J. Dvorak, and V. Bohackova, *Electrochemistry*, Methuen, London, 1966.
4. A. J. Bard and L. R. Faulkner, *Electrochemical Methods, Fundamentals and Applications*, Wiley, New York, 1980.
5. J. Janata, Chemically sensitive field-effect transistors, in: *Solid State Chemical Sensors* (J. Janata and R. J. Huber, eds.), Academic Press, New York, 1985.
6. K. Cammann, Ion-selective bulk membranes as models for biomembranes, in: *Current Topics in Chemistry*, Vol. 128, Springer-Verlag, Berlin, 1985.
7. R. Dohner, D. Wegmann, W. E. Morf, and W. Simon, *Anal. Chem.* 58 (1986) 2589.
8. P. A. Comte and J. Janata, *Anal Chim. Acta* 101 (1978) 247.
9. R. L. Smith and D. C. Scott, *IEEE Trans. Biomed. Eng. BME*-33 (1986) 83.
10. W. E. Morf, *The Principles of Ion-Selective Electrodes and of Membrane Transport*, Elsevier, New York, 1981.
11. J. Koryta and K. Stulik, *Ion-Selective Electrodes*, Cambridge University Press, Cambridge, 1983.
12. R. P. Buck, Electrochemistry of ion-selective electrodes, in: *Chemically Sensitive Electronic Devices* (J. Zemel and P. Bergveld, eds.), Elsevier, New York, 1981.
13. G. Eisenman, in: *Glass Electrodes for Hydrogen and Other Cations*, Chapter 7, Dekker, New York, 1969.
14. M. S. Frant and J. W. Ross, Jr., *Science* 154 (1966) 473.
15. M. L. Iglehart, R. P. Buck, and E. Pungor, *Anal. Chem.* 60 (1988) 290.
16. Xizhong Li, E. M. J. Verpoorte, and D. J. Harrison, *Anal. Chem.* 60 (1988) 493.
17. H. Freiser, Coated wire ion selective electrodes, in: *Ion-Selective Electrodes in Analytical Chemistry* (H. Freiser, ed.), Vol. 2, Chapter 2, Plenum Press, New York, 1980.
18. E. J. Fogt, D. F. Unteraker, M. S. Norenberg, and M. E. Meyerhoff, *Anal. Chem.* 57 (1985) 1995.
19. J. Ruzicka and C. G. Lamm, Anal. Chim. Acta 54 (1971) 1.
20. T. A. Fjeldly and K. Nagy, *Sensors and Actuators* 8 (1985) 261.
21. J. Janata and R. J. Huber, Chemically sensitive field-effect transistors, in: *Ion-Selective Electrodes in Analytical Chemistry* (H. Freiser, ed.), Vol. 2, Plenum Press, New York, 1980.
22. P. Bergveld, *IEEE Trans. Biomed. Eng.* BME-19 (1970) 70.
23. T. Matsuo, M. Esashi, and K. Inuma, Dig. Joint Meet. Tohoku Sect. IEEEJ, 1971.
24. J. R. Sandifer, *Anal. Chem.* 69 (1988) 1553.
25. S. D. Caras, D. Petelenz, and J. Janata, *Anal. Chem.* 57 (1985) 1920.
26. S. D. Caras and J. Janata, Anal. Chem. 57 (1985) 1924.
27. G. G. Guilbault, Enzyme electrodes, in: *Biomedical Investigations, in Medical and Biological Applications of Electrochemical Devices* (J. Koryta, Ed.), Wiley, New York, 1980.
28. M. A. Arnold, Ion-Sel. Electrode Rev. 8 (1986) 85.
29. J. W. Severinghaus, in: *Handbook of Physiology*, Vol. II (W. O. Fenn and H. Rahn, eds.), American Physiological Society, Washington, D.C., 1965.
30. C. S. G. Phillips, *J. Sci. Instrum.* 28 (1951) 342.
31. J. H. Griffith and C. S. G. Phillips, *J. Chem. Soc.* 3446 (1954).
32. I. Lundstrom and C. Svensson, Gas-sensitive metal gate semiconductor devices, in: *Solid State Chemical Sensors* (J. Janata and R. J. Huber, eds.), Academic Press, New York, 1985.
33. M. Josowicz and J. Janata, Suspended gate field-effect transistor, in: *Chemical Sensor Technology*, Vol. 1 (T. Seiyama, ed.), Elsevier, New York, 1988.

34. Y. Miyahara, K. Tsukada, and H. Miyagi, Proc. 6th Sensor Symp. (Jap.), The Institute of Electrical Engineers of Japan, 1986, p. 261.
35. H. Reiss, *J. Phys. Chem.* 89 (1985) 3789.
36. R. A. Bull, F.-R. Fan, and A. J. Bard, *J. Electrochem. Soc.* 131 (1984) 687.
37. M. Salmon, A. Martinez, and K. K. Kanazawa, *J. Electroanal. Chem.* 130 (1981) 181.
38. O. Inganas and I. Lundstrom, *Synth. Metals* 10 (1984) 5.
39. I. Lundstrom and D. Södeberg, *Sensors and Actuators* 2 (1981/82) 105.
40. D. Krey, K. Dobos, and G. Zimmer, *Sensors and Actuators* 3 (1982/83) 169.
41. A. Pelloux, P. Fabry, and P. Durante, *Sensors and Actuators* 7 (1985) 245.
42. N. Yamazoe, J. Hisamoto, N. Miura, and S. Kuwata, *Sensors and Actuators* 12 (1987) 415.
43. W. Weppner, *Sensors and Actuators* 12 (1987) 107.
44. J. Fouletier and E. Siebert, *Ion-Sel. Electrode Rev.* 8 (1986) 133.
45. R. M. Wightman and D. O. Wipf, Ultramicroelectrodes, in: *Electroanalytical Chemistry,* Vol. 15 (A. J. Bard, ed.), Dekker, New York, 1988.
46. M. Fleischmann, S. Pons, D. R. Rolison, and P. P. Schmidt, *Ultramicroelectrodes,* Datatech Systems Inc., Morganton, N.C., 1987.
47. Hsuang-Jung Huang, Peixing He, and L. R. Faulkner, *Anal. Chem.* 58 (1986) 2889.
48. T. Schulmeister and F. Scheller, *Anal. Chim. Acta* 170 (1985) 279.
49. T. Schulmeister and F. Scheller, *Anal. Chim. Acta* 171 (1985) 111.
50. T. Schulmeister, *Anal. Chim. Acta* 198 (1987) 223.
51. A. E. G. Cass, G. Davis, G. D. Francis, H. A. O. Hill, W. J. Aston, J. I. Higgins, E. V. Plotkin, L. D. L. Scott, and A. P. F. Turner, *Anal. Chem.* 56 (1984) 667.
52. Y. Degani and A. Heller, *J. Phys. Chem.* 91 (1987) 1285.
53. M. Umana and J. Waller, *Anal. Chem.* 58 (1986) 2979.
54. W. J. Albery, P. N. Bartlett, M. Bycroft, D. H. Craston, and B. J. Driscoll, *J. Electroanal. Chem.* 218 (1987) 119.
55. C. F. M. Kingdom, *Appl. Microbiol. Biotechnol.* 21 (1985) 176.
56. C. A. Marrese, O. Miyawaki, and L. B. Wingard, Jr., *Anal. Chem.* 59 (1987) 248.
57. D. A. Gough, J. K. Leypoldt, and J. C. Armour, *Diabetes Care* 5 (1982) 190.
58. D. A. Gough, J. Y. Lucisano, and P. H. S. Tse, *Anal. Chem.* 57 (1985) 2351.
59. R. W. Murray, in: *Electroanalytical Chemistry,* Vol. 13 (A. J. Bard, ed.), Dekker, New York, 1984.
60. M. S. Wrighton (ed.), *Interfacial Photoprocesses: Energy Conversion and Synthesis,* ACS Advances in Chemistry Series, Vol. 184, American Chemical Society, Washington, D.C., 1980.
61. G. A. Gerhardt, A. F. Oke, G. Nagy, B. Moghaddam, and R. N. Adams, *Brain Res.* 290 (1984) 390.
62. H. C. Hurrell and H. D. Abruna, *Anal. Chem.* 60 (1988) 254.
63. F. Kreuzer, H. P. Kimmich, and M. Brezina, Polarographic determination of oxygen in biological materials, in: *Medical and Biological Applications of Electrochemical Devices* (J. Koryta, ed.), Wiley, London, 1980.
64. I. Fatt, *Polarographic Oxygen Sensor,* CRC Press, Cleveland, 1976.
65. M. L. Hitchman, *Measurement of Dissolved Oxygen,* Wiley, New York, 1978.
66. Y. Saito, *J. Appl. Physiol.* 23 (1967) 979.
67. V. Vacek, V. Linek, and J. Sinkule, *J. Electrochem. Soc.* 133 (1986) 540.
68. T. Otagawa, S. Zaromb, and J. Stetter, *J. Electrochem. Soc.* 132 (1985) 2951.
69. K. Saji, *J. Electrochem. Soc.* 134 (1987) 2430.
70. E. M. Logothetis, Oxygen sensors for automotive applications, in: *Chemical Sensors* (D. R. Turner, ed.), p. 142, Electrochemical Society, 1987.
71. S. M. Sze, *Physics of Semiconductor Devices,* p. 555, Wiley, New York, 1981.

72. G. Heiland, *Sensors and Actuators* 2 (1982) 343.
73. G. Heiland and D. Kohl, in: *Chemical Sensor Technology*, Vol. 1 (T. Seiyama, ed.), Elsevier, Amsterdam, 1988.
74. W. Göpel, *Prog. Surf. Sci.* 20 (1985) 9.
75. R. M. Warner, Jr., and B. L. Grung, *Transistors*, Wiley, New York, 1983.
76. J. J. McNerney, P. R. Buseck, and R. C. Hanson, *Science* 178 (1972) 612.
77. M. Josowicz, J. Janata, and M. Levy, *J. Electrochem. Soc.* 135 (1988) 112.
78. A. Wilson and R. A. Collins, *Sensors and Actuators* 12 (1987) 389.
79. J. Ghoroghchian, F. Sarfarazi, T. Dibble, J. Cassidy, J. J. Smith, A. Russell, G. Dunmore, M. Fleischmann, and S. Pons, *Anal. Chem.* 58 (1986) 2278.
80. B. Bott and T. A. Jones, *Sensors and Actuators* 5 (1984) 43.
81. R. H. Tredgold, F. C. J. Young, P. Hodge, and A. Hoorfar, *IEEE Proc.* 132, Pt. 1 (1985) 151.
82. S. Bruckenstein and J. S. Symanski, *J. Chem. Soc., Faraday Trans. 1* 82 (1986) 1105.
83. N. Yamazoe and Y. Shimizu, *Sensors and Actuators* 10 (1986) 379.
84. H. Grange, C. Bieth, H. Boucher, and G. Delapierre, *Sensors and Actuators* 12 (1987) 291.
85. M. Hijikigawa, H. Furubayashi, S. Miyoshi, and Y. Inami, Proc. 4th Sensor Symp. (Jap.), Institute of Electrical Engineers of Japan, 1984, p. 135.
86. S. L. Garverick and S. D. Senturia, *IEEE Trans. Electron Devices* ED-29 (1982) 90.

Optical Sensors

The interaction of electromagnetic radiation with matter occurs over a broad range of frequencies and usually in a highly specific manner (Table 5-1). The study and use of these interactions comprise the domain of spectroscopy that provides information ranging from the electronic structure of atoms to the dynamics of polymeric chains. In the most general conventional arrangement the sample (which may be solid, liquid, or gas) is irradiated with monochromatic (single-wavelength) radiation and the extent of the interaction is evaluated from the attenuation of the original radiation or by observing the secondary radiation emitted from the sample (Figure 5-1). The absorption of the primary radiation can also be coupled to other, nonoptical effects, such as the increase in temperature or pressure, or the change in electrical conductivity.

Optical sensors and classical spectroscopic measurements employ the same equipment, but the arrangement of the experiment itself is different. In spectroscopic measurements the sample is generally placed in a well-defined path of the beam and the emerging radiation is captured by the detector. On the other hand, in optical sensors the beam is guided out of the spectrophotometer, allowed to interact with the sample, and then reintroduced into the spectrophotometer in either its primary or secondary form for further processing. The necessity to guide and manipulate the light over some distance limits the frequency range within which optical sensors can be used. Optical waveguides and optical fibers are unavailable for the transmission of radiation at all spectroscopic frequencies. The rapid development of fiber optical sensors parallels (with some delay) the development of optical fibers for communications. At present the frequency range of optical fibers covers well the visible range of the spectrum.

The depository of spectroscopic knowledge forms the basis for the development of optical sensors. In this respect the situation is similar to

Table 5-1. Molecular Spectroscopy by Optical Sensors

Region	Wave number (cm^{-1})	Wavelength (μm)	Transition	Information	Energy ΔE $(kcal\ mol^{-1})$
Far-IR	10^{11}–10^{12}	10–1000	Rotations	Interactions	0.1
IR–Raman	10^{12}–10^{14}	1–10	Vibrations	Functional groups	1
UV–Vis	10^{14}–10^{16}	0.1–1	Electronic shell	Ionization energy	10–100
X-ray	10^{16}–10^{17}	0.01–0.1	Electronic core	Bond dissociation	10^2–10^3

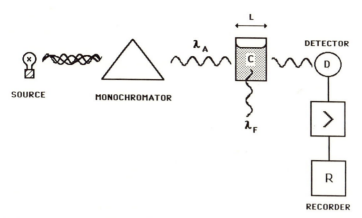

Figure 5-1. General arrangement of a spectroscopic absorption/emission experiment. Cuvette of a defined path length L contains solution of concentration C. The signal from detector D is amplified ($>$) and recorded.

that of electrochemical sensors, which benefit from the experience of general electrochemistry. In this introductory section the basic elements of spectroscopy as they relate to optical sensors will be reviewed first, followed by a brief resumé of the physics of optical waveguides and optical fibers.

5.1. Survey of Optical Spectroscopy[1,2]

Light consists of two sinusoidally oscillating, perpendicularly oriented, electric (E) and magnetic (H) fields (Figure 5-2). They are in phase and move through space with velocity c (2.9979×10^8 m s^{-1} in vacuum). The frequency of oscillation v, wavelength λ, and velocity c are related by

$$\lambda v = c \tag{5-1}$$

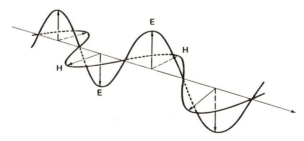

Figure 5-2. Propagation of electric (E) and magnetic (H) vectors in light.

The derived quantity is the wave number v, which is the reciprocal of λ and specifies the number of wavelengths per unit distance (usually cm). The energy contained in light is quantized, each quantum of energy ΔE_λ being called a photon and defined according to the Bohr model as

$$\Delta E_\lambda = hv = hc/\lambda = hcv \tag{5-2}$$

where h is Planck's constant (6.625×10^{-27} erg s). From this expression the energy of the electromagnetic radiation is seen to increase with frequency and wave number (Table 5-1).

When light interacts with matter several processes can take place, sometimes simultaneously. It can be *absorbed*, leading to a decrease in the intensity of the primary beam, or it can be *transmitted* without attenuation. On reaching an interface light can be *reflected* or *refracted*. The last two conditions will be seen as particularly important in guiding light through optical fibers and waveguides. If the direction and/or frequency of the light is changed upon interaction, we talk about *scattering. Fluorescence* and *phosphorescence* are special forms of scattering.

When two or more light beams are considered, the question of the phase relationship between all photons in the two beams, their *coherence*, becomes important. Light can be coherent, in which case the photons are in-phase and interaction between the two beams produces interference fringes with unit visibility. If the light is *noncoherent*, the phase relationship between individual photons is random and the intensities of the two beams add photometrically. The light produced by optical lasers is coherent. When light passes through a condensed phase of oriented dipoles (polarizer), or when it is reflected, it becomes *polarized* with the electric vector parallel to the medium dipoles or, in the case of reflection, oriented in the azimuthal plane. Polarized light will be attenuated by the second polarizer rotated 90° with respect to the primary polarization.

As the frequency is scanned the degree of absorption of the interacting light varies. The relationship between the relative absorption and the wavelength is called the *absorption spectrum* and is characteristic of the properties of the medium. Examination of the energy belonging to the different parts of the spectrum (Table 5-1) shows that these interactions will involve different properties of the molecule. The domain of optical sensors is in the ultraviolet and visible part of the spectrum, which means that the interactions will be limited more or less to the transitions at electronic levels (Figure 5-3) corresponding to changes in the free energy on the order of tens of kcal mol^{-1}.

When a molecule absorbs a photon possessing an exact amount of energy, it moves from its electronic ground state to its first excited state. Within each electronic state the molecule can occupy different vibrational

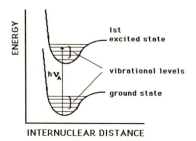

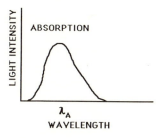

Figure 5-3. Electronic transitions caused by interaction with light and the corresponding absorption spectrum.

levels, which are separated from each other by energy differences less than 1 kcal mol^{-1}. Figure 5-3 is not drawn to scale; the difference between the two electronic levels is much larger than the difference between the individual vibrational levels. The excited molecule can lose its energy dissipatively, by rapid weak collisions with its neighboring molecules. In this case the energy of the molecule decreases within the excited state toward its lowest vibrational level. If the molecule can remain in the lowest vibrational level of the excited state long enough, it can make the transition to the ground state either by a strong collision with another molecule or radiatively. In the latter case the energy loss occurs in a quantized fashion and the emitted light again has a characteristic *fluorescent* frequency (Figure 5-4). In general, molecules in gas and in liquid phase do not have a fixed orientation with respect to the exciting light. Therefore, the emission of fluorescence occurs in all directions. The absorption and emission processes are symmetrical, so the corresponding spectra are also symmetrical in appearance.

The emitting molecule is called a *fluor*. Energy dissipation by strong collisions is clearly a competing process, which limits the efficiency of the fluorescence. Different molecules possess different ability to dissipate energy from the fluor, partly because of their concentration but also because of their stereospecific fit to the excited molecule. Molecules which can

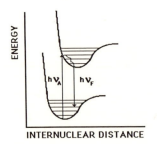

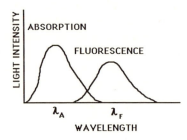

Figure 5-4. Absorbed energy may be dissipated nonradiatively to the lowest vibrational level of the first excited state. Transition from that level to any vibrational level of the ground state is accompanied by radiation of characteristic frequency—fluorescence.

dissipate energy from an excited molecule by a thermal process even at low concentrations are called *quenchers*. The fluorescence intensity I_F depends on the *quantum efficiency Q* given by

$$Q = \text{(photons emitted)}/\text{(photons absorbed)} \qquad (5\text{-}3)$$

It also depends on the concentration of the fluor C, on its molar absorptivity ε, and on the pathlength L. Usually only a fraction of the emitted light is captured by the detector, hence a collection factor γ is also introduced. The final expression is therefore

$$I_F = \gamma Q \varepsilon c L \qquad (5\text{-}4)$$

Fluorescence can also be treated from the viewpoint of its lifetime. Both specific and nonspecific quenching is a competitive process, so the *fluorescence lifetime* ($1/e$ value of the original intensity) is a characteristic parameter that can be employed to describe the concentration of both the fluor and the quencher. Typical fluorescence lifetimes are in nanoseconds. Another effect that could be used for chemical sensing is *photobleaching*, which is the degradation of the fluor due to irreversible chemical reactions of the molecule in its excited state.

There is an inherent similarity between the spectrum and an electro-chemical current–voltage curve that is important from the standpoint of chemical selectivity. In both cases the x-axis is directly related to the energy of interaction. In the electrochemical case this energy corresponds to trans-fer of electrons from the solution species to/from the electrode, namely, the standard potential. In optical interaction it is related to the difference in energy of two levels of the molecule. There is, however, a fundamental difference which makes the optical selectivity far more important than that derived from the standard potential of an electrochemical reaction. In electrochemistry, at a given applied potential all electrochemical reactions possessing equal or lower energy will take place. This means that the selec-tivity can be defined in only a very limited way by the choice of electrode potential. On the other hand, optical transitions are limited on both sides: if not enough energy is applied or if too much energy is applied, the transition (and therefore the interaction) will not take place. Consequently, many more individual absorption processes can be accommodated on the frequency (energy) axis. Their actual number is indirectly proportional to the linewidth. According to equation (5-2) there is a sharply defined amount of energy associated with the transition which would correspond to a single spectral line. Such a line spectrum is observed, for example, for atomic vapors. On the other hand, spectral lines of more complicated molecules, even in the gas phase, are broader. This is because transition between two electronic states is complicated by the existence of multiple vibrational levels within each state. Furthermore, in the condensed phase these vibrational levels are strongly affected by interactions with surround-ing molecules. The second reason for line broadening is that electronic (absorption) transitions may take place between the ground state and the first or higher excited states. However, because fluorescence always takes place between the first excited state and the ground state, the fluorescent spectra are usually sharper than the corresponding absorption spectra. Another reason for line broadening is the change in the internuclear dis-tance between the atoms during the event of absorption of the quantum by the molecule (the Franck–Condon effect). This effect has no direct implica-tion for chemical sensors, except that line broadening in general represents a certain loss of chemical selectivity.

Phosphorescence is similar to fluorescence but based on a slightly different mechanism. The transition to the first excited (singlet) state on absorption of a quantum of light is the same. If there is a triplet state (with two unpaired electrons) lying close below the first excited state, a non-radiative singlet–triplet transition can take place (Figure 5-5). This transition has low probability because it is "forbidden" by the rule which states that a change in the multiplicity of electron spins is not allowed. Radiative decay from the triplet state to the ground state, analogous to

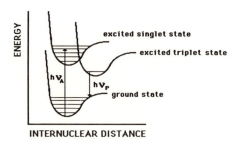

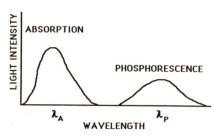

Figure 5-5. Absorbed energy may be dissipated nonradiatively and the molecule makes a transition from the excited singlet to the excited triplet state. The ensuing transition to the ground state results in the emission of phosphorescence.

fluorescent emission, then takes place. Multiple rearrangements must occur during this process, hence phosphorescence has a much longer lifetime (milliseconds) than fluorescence.

The last optical process which offers sensing possibilities is chemiluminescence. In this case the excited state is created by a chemical reaction. The excited state then emits light according to the scheme

$$A + B \Rightarrow C^* \Rightarrow C + h\nu$$

There are numerous electrochemical reactions in which electro-luminescence occurs as a result of chemical reactions that follow the initial charge-transfer reaction. Chemical sensors based on the electroluminescent effect have been proposed. The transition which fits the definition of equation (5-2) can be called an "allowed" transition. When the sample is irradiated with light of higher frequency (higher energy) than would correspond to any electronic transition, the majority of the light passes through the sample without attenuation but a small fraction is scattered, giving rise to very faint lines corresponding to transitions within vibrational levels (Figure 5-6). These lines belong to the *Raman spectrum* and are observed at right angles to the incident beam. They are "not allowed"

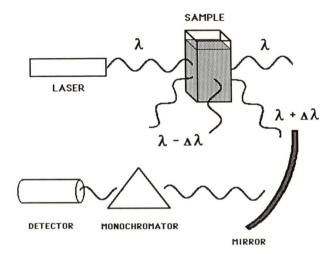

Figure 5-6. Scattering of incident radiation of wavelength λ. Stokes and anti-Stokes scattering results in the formation of symmetrical lines of wavelength $\lambda + \Delta\lambda$ and $\lambda - \Delta\lambda$.

in the sense of the Bohr model if it is taken exactly. However, if we realize that in quantum mechanics "allowed" means "with higher probability" and "not allowed" means "with lower probability," then the existence of these transitions can be rationalized. Weak Raman lines (Figure 5-7) shifted toward lower frequencies are called Stokes lines. Situated symmetrically with respect to the Rayleigh line we find a somewhat less intense series of anti-Stokes lines, which correspond to transfer of energy from the molecule to the scattered light. Closer to the central line lie Brillouin lines, which are caused by movement of the scatterer within the time interval of the interaction. Like the Franck–Condon effect these lines do not seem to be relevant to chemical sensors at present.

The difference in frequency between the Rayleigh line and one of the Raman lines corresponds directly to the energy difference of the transition between two vibrational levels in line with equation (5-2). The Raman

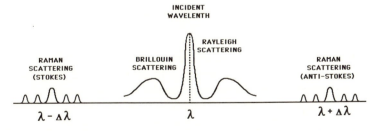

Figure 5-7. Rayleigh, Raman, and Brillouin lines produced by incident radiation.

effect thus provides information which is normally the domain of infrared spectroscopy. Yet, any sufficiently strong source which emits in the visible part of the spectrum, such as a laser, can be used as the exciting radiation. This is particularly important from the sensing point of view, because optical fibers and waveguides for the transmission of a broad spectrum of infrared light do not exist and normal-mode infrared spectroscopy still lies beyond the reach of common optical sensors.

5.1.1. The Lambert–Beers Law

Most optical sensing processes rely on the absorption of light as the first step. There is a quantitative relationship between the concentration of the absorber, molar absorptivity, and the optical path length. For its derivation it is convenient to invoke the "particulate" character of light, which says that a photon is an optical particle carrying a discrete amount of energy—a *quantum*. The intensity of a beam of light of unit cross-sectional area is

$$I = N \tag{5-5}$$

where N is the number of photons per unit area. When light passes through an absorber of elementary volume $dV = A\,dx$, where A is the unit area, the number of photons absorbed equals dN, leading to a decrease in light intensity,

$$dN = -dI \tag{5-6}$$

The probability P that absorption of a photon will take place is proportional to the number of absorbing molecules n present in the elementary volume,

$$P = dN/N = an = aC(A\,dx) \tag{5-7}$$

where C is concentration in arbitrary units. Then

$$\int_{I_0}^{I} -dI/I = aC \int_{x}^{x+\Delta x} dx \tag{5-8}$$

I_0 being the intensity of the incident beam and I the final intensity. Integration of equation (5-8) over the entire path length L and conversion to a base-10 logarithmic scale yields the Lambert–Beers law, which relates absorbance A to concentration,

$$A = -\log I/I_0 = \varepsilon CL \tag{5-9}$$

The proportionality constant ε is called the absorption (or extinction) coefficient or, when C is expressed as molar concentration, it is called *molar absorptivity* (or the molar extinction coefficient). From this equation it is seen that the variation in the attenuation of the initial beam is equally affected by changes in the optical path length and changes in concentration. For a normal spectrophotometric experiment the constant optical path L is defined by the spacing of the transparent "cuvette" windows. The equivalent rigid definition of L does not usually exist in fiber optical sensors.

Molar absorptivity is a function of wavelength and is an additive property. This means that for samples containing multiple absorbing species within the same optical path and at the same wavelength, the absorbance A is given by

$$\log I_0/I = A = x(\varepsilon_1 C_1 + \varepsilon_2 C_2 + \cdots + \varepsilon_y C_y) \qquad (5\text{-}10)$$

where the εC quantities are the contributions of the individual species to the total attenuation. The shapes of the absorption spectra of different species vary (i.e., the ε quantities are different functions of λ), so equation (5-10) serves as a powerful tool in the quantitative spectrophotometry of complex mixtures.

It may be useful to pause for a moment and review the factors which affect the selectivity of optical sensors. First of all the resonant nature of the interaction between light and the absorber is one of the most important factors in achieving selectivity. It is further enhanced by the geometrical effects which distinguish the spatial distribution of the fluorescent radiation from the linear character of the primary absorption step. Additional selectivity accrues from the use of polarized light and from the time-resolved fluorescent and phosphorescent measurements. On the other hand, some selectivity is lost due to line broadening. In summary, the selectivity inherent in the transduction step in optical sensors is far higher than in any other type of sensor. Obviously, like any other sensors, optical sensors utilize also all the benefits of selectivity originating from the chemically selective layer.

5.2. Waveguides and Optical Fibers

A typical optical sensor consists of different modules, such as a source, monochromator, chopper, sample space, detector, and so on. Light is transmitted between these modules by cylindrical optical fibers or by planar waveguides according to the following scheme:

W   A   V   E   G   U   I   D   E

light source → (monochromator) → sample → (monochromator)   ┐
                                                                                      ¦
                                                                                      ¦
                                                                                      ¦
                                                                   WAVE GUIDE ¦
                                                                                      ¦
                              display electronics ← detector   ⌐

In this path the beam interacts continuously with interfaces between media of different optical density (refractive index). The laws which govern its behavior at these boundaries are most relevant for the operation of optical sensors.[3]

5.2.1. Reflection and Refraction[4]

When light arrives through an optically dense medium of refractive index n_1 at its boundary with an optically rarer medium of refractive index n_2 (i.e., $n_1 > n_2$) it can be either *reflected* back to phase 1 or *refracted* (transmitted through) to phase 2. Which of these two conditions applies is determined by the *angle of incidence* Θ_1 at the interface (Figure 5-8). The speed of light is lower in the medium of higher optical density. The Maxwell equations require that the phase of the transmitted light Θ_2 and its frequency be preserved on passing through the interface. Velocity and frequency are related to wavelength by equation (5-1), so the angle of the transmitted light must change in order to satisfy the condition of phase matching. This is why objects submerged in water appear to be broken but have the same color as the original object. Naturally, the angle of reflected light Θ_1 is unchanged and equals the angle of the incident beam.

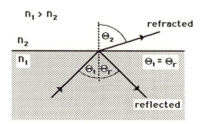

Figure 5-8. Reflection and refraction of light at the interface between two media with different refractive indices.

The geometrical relationships for refracted light are given by Snell's law

$$n_1 \sin \Theta_1 = n_2 \sin \Theta_2 \tag{5-11}$$

When the angle of incidence is increased beyond certain value, the condition of phase matching can no longer be satisfied and all the light is transmitted along the interface (i.e., $\Theta_2 = 90°$). The angle at which this happens is called the *critical angle* Θ_c for which equation (5-11) reduces to

$$\sin \Theta_c = n_2/n_1 \tag{5-12}$$

For the air–water interface Θ_c is 43.75°. For values $\Theta_1 > \Theta_c$ all the light is reflected to the optically dense phase. This is called the condition of *total internal reflection* (TIR).

In order to propagate an electromagnetic wave along the interface or to reflect it totally, there must be a zero net flux of energy into the optically rare medium (n_2). However, there is a finite decaying electric field across the interface which extends some distance x into the rarer medium. The intensity of this field is given by

$$E_2 = E_0 \exp\left[\frac{-i2\pi n_2}{\lambda}(x \sin \Theta_2 + z \cos \Theta_2)\right] \tag{5-13}$$

When $\Theta_1 = \Theta_c$ and $\Theta_2 = 90°$, equation (5-13) reduces to

$$E_2 = E_0 \exp \frac{-i2\pi n_2 x}{\lambda} \tag{5-14}$$

For values of Θ_1 greater than Θ_c, the combination of equations (5-11) and (5-12) produces a seemingly paradoxical result: $\sin \Theta_2 > 1$. It can be accepted if we realize that the electric vector is a complex number. Therefore, from equation (5-11) and using the equality $\sin^2 \Theta + \cos^2 \Theta = 1$, we obtain

$$\cos \Theta_2 = [1 - (n_1/n_2)^2 \sin^2 \Theta_1]^{1/2} \tag{5-15}$$

For $\Theta_1 > \Theta_c$ the term within the root sign is negative. Therefore

$$\cos \Theta_2 = \pm i[(n_1/n_2)^2 \sin^2 \Theta_1 - 1]^{1/2} \tag{5-16}$$

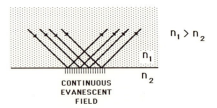

CONTINUOUS
EVANESCENT
FIELD

Figure 5-9. Origin of the continuous evanescent field at the interface.

Equations (5-13) and (5-16) can be combined to yield the value of the electric field which penetrates the optically rarer medium (n_2) at the point of reflection,

$$E_2 = E_0 \exp\left\{\frac{-i2\pi}{\lambda}(xn_1 \sin \Theta_1 \pm iz[(n_1/n_2)^2 \sin^2 \Theta_1 - 1]^{1/2}\right\} \quad (5\text{-}17)$$

or

$$E_2 = E_0 \exp\left[\frac{-i2\pi}{\lambda}xn_1 \sin \Theta_1\right]\exp\left\{\pm\frac{2\pi z}{\lambda}[(n_1/n_2)^2 \sin^2 \Theta_1 - 1]^{1/2}\right\}$$

$$(5\text{-}18)$$

The first exponential term in equation (5-18) is responsible for the shift of the reflected wave in the direction of its propagation (Figure 5-9). The second term accounts for the exponential decay of the field intensity in the direction normal to the interface.

The reflected beam combines with the incident beam to form a standing electromagnetic wave at the interface (Figure 5-10). The electric

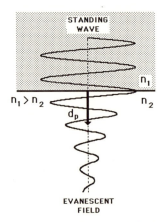

STANDING
WAVE

$n_1 > n_2$

n_1

n_2

d_p

EVANESCENT
FIELD

Figure 5-10. Standing wave and exponential decay of the intensity of the evanescent field according to equation (5-19).

field which penetrates the optically rare medium is called the *evanescent field*. The depth of its penetration d_p is then defined as the distance at which the initial intensity decays to its $1/e$ value. Thus from equation (5-18) d_p is defined as

$$d_p = \frac{\lambda}{2\pi(n_1^2 \sin^2 \Theta_1 - n_2^2)^{1/2}} \tag{5-19}$$

It is important to note that the depth to which the evanescent wave penetrates depends on several parameters, but above all on the wavelength of the incident beam. This consideration will be important in sensing applications in which the interaction of a chemical species with this field gives rise to an analytical signal. For common materials and visible light, d_p is typically between 1000 and 2000 Å.

There is another way in which we can look at the phenomenon of optical absorption than that described within the framework of the Lambert–Beers law. The attenuation of passing electromagnetic radiation by the absorbing medium can be seen as the result of any interaction of this radiation with the medium. The ordinary refractive index n is part of the complex index of refraction $\tilde{n}$, which can be written as

$$\tilde{n} = n - ik \tag{5-20}$$

Constant k is the wavelength-dependent absorption index, which is related to molar absorptivity:

$$k(\lambda) = \lambda\varepsilon(\lambda)/4\pi \tag{5-21}$$

Therefore, the Lambert–Beers law formulated in terms of a complex index is

$$\log I/I_0 = 4\pi kx/\lambda \tag{5-22}$$

5.2.2. Multiple Total Internal Reflection

The above phenomenon of the reflection of light has two important implications for chemical sensing. First, it confines the radiation into one phase and, as such, forms the basis of propagation of light through waveguides and fibers.[5] The second factor is the existence of the evanescent field, which represents an experimental space in which the chemical interactions can take place.

When a beam of light is introduced into a thick planar waveguide (Figure 5-11) surrounded by a homogeneous medium of lower refractive index, light bounces from the opposite interfaces and propagates in zigzag

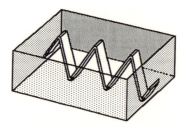

Figure 5-11. Multiple reflections in a flat optical waveguide.

fashion through the waveguide. Naturally, at each point of reflection there is an evanescent field. If there is no interaction within this field and/or within the waveguide, the light will pass without attenuation. If the thickness is much greater than the wavelength of light, the boundary conditions at the two reflecting planes can be satisfied independently of each other. Let us now decrease the thickness and "squeeze" the light. We see that the points of reflection move closer to each other. As they begin to overlap the reflected beams begin to interfere. This produces another set of conditions, which must be satisfied from the viewpoints of both the single reflection and the constructive interference. As a result of this additional constraint a finite number of reflection angles exists for each wavelength at which the conditions allow propagation by multiple total reflection. These constructive, allowed patterns are called *guided modes* of the waveguide (Figure 5-12). The thinner the waveguide and the smaller the difference between the refractive indices, clearly the fewer the modes. When the waveguide operates in a multiple reflection mode, light no longer exists as an individual beam but rather appears as a continuous streak.

The second consequence of increasing the number of multiple reflections is the appearance of the continuous evanescent field, which results from overlap of the individual evanescent points. This may serve in sensing applications as a large "active area" in which the sensing interaction can take place.

An essentially identical situation exists for the multiple reflection in optical fibers. In this case the boundary conditions for the multiple reflections and considerations governing the existence and number of guided

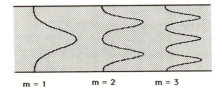

m = 1   m = 2   m = 3

Figure 5-12. Guided modes in an optical fiber.

modes are formulated in cylindrical coordinates. The actual propagation path is spiral as opposed to zigzag in planar waveguides, as can be visualized by using coherent laser light. For a noncoherent source the light propagated by a fiber again appears as a continuum of intensity. The number of modes M in a fiber of radius r (both r and λ are in μm) is then given as

$$M = (2\pi r)^2 (n_{co}^2 - n_{cl}^2)/2\lambda^2 \tag{5-23}$$

where n_{co} and n_{cl} are the refractive indices of the fiber core and cladding, respectively. A typical fiber of $d < 10$ μm supports one mode (a single-mode fiber), while thousands of modes can be accommodated in fibers which are approximately 100 μm in diameter.

A typical construction of an optical fiber is shown in Figure 5-13. The optical core of refractive index n_{co} (a typical material is silica, which has $n_{co} = 1.6$) is surrounded by an optical insulator, called cladding, which has lower refractive index n_{cl} (a typical value is 1.5). The outside coating on the fiber is there for protection and has no special optical function.

In order for the light to be coupled, it must enter the end of the fiber within a certain cone whose angle is related to the critical angle for reflection at the core–cladding interface. Likewise, light leaving the end of the fiber spreads only over the same solid angle. For a rational design of an optical sensor it is important to know the shape of this cone. The geometry of the end of the fiber is shown in Figure 5-14. We are trying to determine the angle Θ_0 which defines the *cone of acceptance*. The primary consideration is the propagation of light within the fiber for which the critical angle is defined by equation (5-12). Refraction at the interface between the end of the fiber and the outside medium (n_0) is governed by Snell's law [equation (5-11)]

$$n_0 \sin \Theta_0 = n_1 \sin(90 - \Theta_c) \tag{5-24}$$

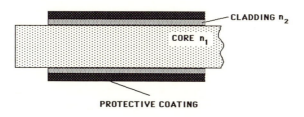

CLADDING n_2

CORE n_1

PROTECTIVE COATING

$n_1 > n_2$

Figure 5-13. Cross-section of an optical fiber.

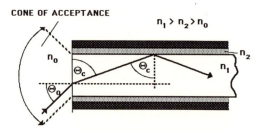

Figure 5-14. The cone of acceptance of an optical fiber.

With the aid of the relationship $\sin(\alpha - \beta) = \sin \alpha \cos \beta - \cos \alpha \sin \beta$, we obtain

$$n_0 \sin \Theta_0 = n_1 \cos \Theta_c \tag{5-25}$$

because

$$n_1^2 \cos^2 \Theta_c + n_1^2 \sin^2 \Theta_c = n_1^2 \tag{5-26}$$

and

$$n_1^2 \sin^2 \Theta_c = n_2^2 \sin^2 90 = n_2^2 \tag{5-27}$$

Equations (5-25)–(5-27) yield

$$n_0^2 \sin^2 \Theta_0 = n_1^2 - n_2^2 \tag{5-28}$$

or

$$NA = (n_1^2 - n_2^2)^{1/2}/n_0 \tag{5-29}$$

where the number NA is called the *numerical aperture* of the fiber and defines the cone of acceptance in terms of refractive indices alone. The definition in terms of the "half-angle" Θ_0 is possible from equation (5-28).

5.2.3. Diffuse Reflection

When the angle of the incident beam unequivocally determines the angle of the reflected beam, we talk about specular reflectance. It occurs at optically smooth interfaces where the amount of scatter is minimal. On the other hand, on rough and granular surfaces the light is scattered over a wide solid angle. It still contains information about the composition of the reflector, but the relationship between the attenuation and concentration is less explicit and, furthermore, it is spatially distributed.

Various models have been proposed which, with varying degrees of success, quantify the reflectance and concentration. Best known is the

Kubelka–Munk model,[6] which is sometimes called "the Lambert–Beers law of reflectance spectroscopy." Function $f(R)$ relates the reflectance R to the absorption (εC) and scattering (S) constants,

$$f(R) = (1 - R)^2/2R = \varepsilon C/S \qquad (5\text{-}30)$$

This relationship works well for many different types of scatterer and for low concentrations of the absorber. The best results have been obtained for mixtures of absorbing and nonabsorbing powders (Figure 5-15) and for an adsorbed absorber on a colorless solid. Fair agreement has been obtained also for the evaluation of colored zones on paper chromatograms. Better results for this type of sample have been obtained using the modified formula[7]

$$R_0/R = (1 + \beta)^2/(1 + \beta/Q) \qquad (5\text{-}31)$$

In this case the value of the reflectance for a pure scatterer, R_0, is used to normalize the measurement. Parameters β and Q are defined in terms of empirical parameters related to reflection (r), transmission (t), and absorption (μ),

$$\beta = \mu/(r + t) \qquad \text{and} \qquad Q = 1 + t/(r + t) \qquad (5\text{-}32)$$

Hence for a highly reflecting and minimally transmitting sample, $Q = 1$ and equation (5-31) simplifies to

$$R_0/R = 1 + \beta \qquad (5\text{-}33)$$

The empirical parameters r, t, and μ are employed to adjust the equation to the morphology of the sample.

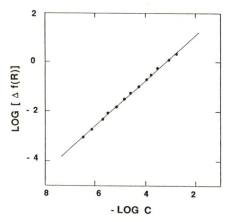

Figure 5-15. Diffuse reflectance at 565 nm from a solid solution of $KMnO_4$ in $KClO_4$ (points). The solid line is calculated according to equation (5-30). (Reprinted from Ref. 7 with permission of the Society for Applied Spectroscopy.)

5.3. Optical Sensors

Optical fibers and waveguides can transmit light over large distances and with minimal loss of intensity. As long as the bends in the fiber are not too sharp (the radius of the bend must be much greater than the wavelength), the direction of propagation can be changed easily. Thus, light can be delivered to and from the active sensing area even if this area is situated far from the source and the detection system. This makes optical sensors particularly attractive for remote sensing and for applications where the use of electricity may be hazardous. There are many different forms of optical sensor that differ from each other in the details of their design. In this section we shall try to pinpoint the essential elements of various implementations of optical sensors in which optical fibers and waveguides are used to pipe light. One generalization should be stated initially. From the spectroscopic point of view there is nothing fundamentally new in what optical sensors do and what could not be done by performing the same experiment in a conventional arrangement with a well-designed spectrophotometer. The convenience of remote sensing, electrical safety, the monolithic character of the optical probe, and all the other real and fictional benefits usually quoted in favor of optical sensors, are introduced the expense of decreased performance as compared to analogous conventional spectroscopic measurement.

There is a great degree of selectivity inherent in the transduction part of optical sensors given by the choice of wavelength(s), polarization, etc. Nevertheless, like in any other type of sensor, a chemically selective layer provides the principle source of selectivity through the specific interactions with the determinand. These interactions are described in terms of equilibria, which must be expressed in activities *and not concentrations* as is often done. It is in the very nature of the interaction between the electromagnetic radiation and matter that it takes place on atomic and molecular levels. In other words, light counts molecules and therefore provides information about *concentration*. Interactions of the molecules among themselves and with their environment (which accounts for their *activity*) usually appears only as secondary effects: shift of the absorption and/or emission peak, line broadening, change of fluorescence decay time, etc. Furthermore, there are no general quantitative rules connecting these secondary effects with the activity coefficient of the molecule in the given environment. Thus, there is a fundamental discrepancy between the transduction mechanism and the origin of selectivity provided by the chemical layer which is inherent in optical sensors.*

---

* The same inherent problem exists also in mass sensors, which also count the number (weight) of added species. However, because their domain is in gas-phase applications problems with fugacity are much less severe than those with the activity in the liquid phase.

We now focus on the selective layer and how it is incorporated in optical sensors. Unless there is a specific reason we shall use the term "optical fiber" in a general sense meaning both optical fibers and optical waveguides. The selective layer can be placed on the fiber in several different ways, depending on the type of interaction used.[8,9] The simplest arrangement for absorption studies involves two fibers facing each other across a gap, which is filled with the selective layer. One fiber delivers and the other collects the light (Figure 5-16a) with some acceptable efficiency. Here the gap dimensions define the optical path length required by the Lambert–Beers law [equation (5-9)]. Another possibility is to use two parallel fibers and a mirror, which reflects the light into the collection fiber. The detection limit is improved, because the light traverses the selective layer twice. A similar arrangement (Figure 5-16b) uses only one *bifurcated* fiber, which transmits both the initial and final intensity. The simplification at the sensing end leads to complications at the instrumentation end, where the two beams must be separated by mirrors, beam splitters, and other necessary hardware and piped to the appropriate parts of the spectrophotometer. The reflecting surface in these two implementations can be a

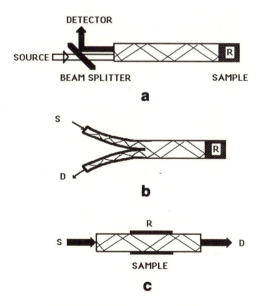

Figure 5-16. Three most common types of arrangement of an optical sensor, where R is the reagent, and S and D are the source and detector, respectively.

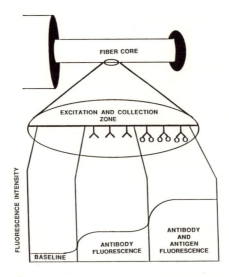

Figure 5-17. Concept of an evanescent wave sensor proposed for monitoring an immunocheemical reaction. Either the antigen or the antibody could be immobilized on the fiber.

mirror or a scattering surface (b). In any case the sensor has the appearance of a monolithic probe (i.e., a "dip-stick" probe*). Optical sensors based on absorption, fluorescence, phosphorescence, and luminiscence can employ these two configurations.

The second possibility is to make use of the continuous evanescent field which exists at the surface of an optical fiber and place the selective layer within this field (Fig. 5-16c). For such an arrangement the protective coating and cladding are replaced[10] by the selective layer in a short segment of the fiber (Figure 5-17). The conventional counterpart is internal reflection spectroscopy in all its variants.[11] The attenuation of primary radiation in an absorption experiment takes place through the imaginary part of the refractive index [equations (5-20) and (5-21)]. The spatial distribution of this interaction depends on the depth of penetration d_p of the evanescent field [equation (5-19)]. How many molecules of the absorber are present within this evanescent field in turn depends on the nature of the adsorption process itself. A quick examination of these equations shows that the "optical path" in evanescent measurements is not constant but rather depends on the wavelength, the refractive indices of all the materials involved, and the geometry of the arrangement. Owing to these complicated interrelated processes, the concentration dependence of

---

* It should be noted here that the description "dip-stick" coupled with other adjectives such as "disposable," "integrated," "solid-state," etc. makes any sensor virtually irresistible to the prospective investor.

the signal is never calculated in an absolute sense but rather established as an empirical calibration curve for a given standard set of experimental conditions.

A particularly advantageous approach to evanescent wave sensing is the measurement of fluorescence. This can be done in two ways. In both cases the primary light is delivered to the selective layer zone through the fiber, and the fluorescence is measured either with the aid of another fiber (Figure 5-18a) or it is coupled back to the same fiber (Figure 5-18b) and the longer-wavelength fluorescence is separated from the exciting radiation with the help of a monochromator at the instrumentation end. The technique of total internal reflectance fluorescence (TIRF) has been applied to the study of protein adsorption using attached fluorescent labels[12] or by

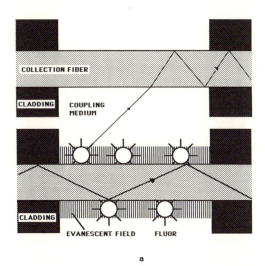

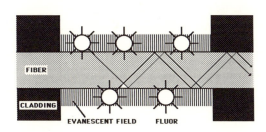

Figure 5-18. Two arrangements for a fluorescent sensor: (a) The excitation and collection are done by two separate fibers. Different geometrical configurations of these two fibers are possible. (b) Both excitation and collection are done by the same fiber.

directly using the intrinsic fluorescence of the absorbing molecules.[13] Both alternatives offer the advantage of rejecting background fluorescence from the bulk of the sample; internal coupling of the fluorescent light has the advantage of signal enhancement due to the higher reported collection efficiency when the fluorescence is excited within the evanescent field.

The parameters which most affect the choice of placement of the chemically sensitive layer are the optical properties of this layer and of the sample itself. If they are optically transparent, except for the absorber, any of the above modes can be used. On the other hand, if the sample is opaque or highly colored then the evanescent wave or reflectance may have an advantage over direct transmission. In any event the low optical transparency of the sample always poses problems for optical sensing.

5.3.1. Ionic Sensors

Optical sensors for ions employ indicators, which exist in two different color forms depending on whether or not the determinand (ligand) is bound to them. The use of colored indicators is one of the oldest principles of analytical chemistry and is used extensively in both direct spectroscopy and visual titrations. The indicator is usually confined to the surface of the optical sensor or immobilized in an adjacent layer. In that sense the oldest and most widespread optical sensor is a pH indicator paper (litmus paper), which is commonly used for rapid and convenient semiquantitative estimate of the pH of solutions. Its hi-tech counterpart is a pH optrode (which intentionally rhymes with pH electrode) that essentially does the same thing.

The operation principles of optical sensors are common for all ions. They will be discussed in the framework of the optrode for the most ubiquitous of all—hydrogen ion and measurement of pH. The primary chemical interaction governing the operation of the pH optrode is the acid–base equilibrium of the indicator HA,

$$HA + H_2O = H^+ + A^-$$

Its dissociation constant is defined by the following equations:

$$K = \frac{a_H C_A}{a_{H_2O} C_{HA}} \frac{f_A}{f_{HA}} = K_a \frac{f_A}{f_{HA} f_{H_2O}} \tag{5-34}$$

In logarithmic form it is known as the Henderson–Hasselbalch equation,

$$pH = pK + \log(C_A/C_{HA}) + \log(f_A f_{HA}) - \log a_{H_2O} \tag{5-35}$$

where C_A and C_{HA} are the concentrations of the indicator conjugate base and conjugate acid, respectively, one or both of which are measured optically. If we assume that a hundredfold change in concentration of the absorbing molecule can be conveniently measured (i.e., the dynamic range pertaining to one K equals $\pm$ one decade), then one indicator can cover two units of pH. By selecting a series of indicators with suitably spaced values of their dissociation constants, an optrode with broad dynamic range can be obtained.

Nearly all papers describing the pH optrode ignore the third and fourth terms in equation (5-35). This can be tolerated for very dilute aqueous solutions where the activity coefficients tend to unity, their ratio converges even faster with dilution, and water is in a larger molar excess. Unfortunately, conditions in most real measuring situations are such that these terms can rarely be ignored. Various ways of dealing with this problem have been suggested.

It is important to realize that electrochemical measurements face a similar problem, which has its origin in the thermodynamical impossibility of measuring the activity of a single ion, including pH. In electrochemistry this problem is usually approached by using electrochemical cells with liquid junctions (cf. Section 4.1). However, a liquid junction separating two dissimilar solutions is a nonequilibrium element which leads to the *operational definition* of pH (X) of an unknown solution,

$$\text{pH (X)} = \text{pH (S)} + [\text{EMF (X)} - \text{EMF (S)}]\, F/2.303\, RT + \varDelta E_j F/2.303\, RT$$

$$(5\text{-}36)$$

where S refers to the standard solution and X to the sample. It is normally assumed that the change in liquid-junction potential $\varDelta E_j$ is negligible, i.e., that it can be calibrated out. It is this nonthermodynamical assumption of the invariability of the liquid-junction potential that represents the most serious flaw in the practical electrochemical determination of pH. For-tunately, it is almost always possible to design a liquid junction such that it is suitable for the given application and that the error in the estimate of pH due to the change in E_j is minimal. In any case, the pH electrode itself does respond to the activity of a hydrated hydrogen ion.

The effect of the environment on the dissociation equilibrium of the indicator used in an optical sensor is through the ratio of the activity coef-ficients of its two forms [equation (5-35)]. These include the change in ionic strength, and specific interactions due to, e.g., proteins, adsorption effects, etc. Attempts to account for some of these interactions on the basis of classical Debye–Huckel theory have been made and an optrode has been constructed for the determination of *ionic strength*.[14] This approach works well for a uni-univalent electrolyte up to ionic strength $I < 0.1$ M. For higher ionic strengths and/or multivalent ions, one must take into account[15] specific interactions which involve the ionic sizes of all ions present in the medium and the type of indicator used.

The effect of proteins on the dissociation constant of indicators, the so-called "protein error," has been known to exist for a long time. It is defined as the difference between the colorimetrically determined values of pH in the presence and absence of the proteins in solution whose pH has been adjusted (electrochemically) to its original value. Depending on the type of proteins and their concentration and on the type of indicator, it can be as high as 0.8 units of pH. Again, it originates in the specific interactions between the indicator and proteins.

Several attempts have been made to overcome these problems. It follows from the above discussions that the relative contribution of the third and fourth terms in equation (5-35) to the observed value of pH depends on both specific and nonspecific interactions. That contribution, in turn, depends on the type of indicator used and on the type and concentrations of all ions in the sample. Thus, in one approach[16] two different types of indicator (hydroxypyrenetrisulfonic acid and B-methylumbelliferone) have been used and measured at two different wavelengths. Since both the specific and nonspecific interaction terms for these two indicators are very different, their dependence on ionic strength is also sufficiently different to yield the two sets of equations necessary to solve for both pH and ionic strength. The problem is that those interaction terms are valid only for one specific type of ion (both the charge and chemical types) and if the general composition of the sample changes, these equations are no longer valid.

Another approach[14] has been to place the indicator in a highly charged environment by surrounding it with quaternary ammonium groups in the hope of making any variations in the ionic strength of the sample negligible. This, however, creates another kind of problem. The most fundamental difference between optical and electrochemical pH sensors is that, in the latter, the signal originates from bulk–bulk interactions while in optical sensors the signal originates from bulk–surface interaction. The color indicator is confined to the surface of the optical fiber and the measured ion must equilibrate between the bulk of the sample and this surface region. Thus equation (5-35) must be written in *surface activities*. In other words, the dissociation equilibrium is based on the conditions as they exist at the sensor interphase (surface and adjacent region) and the above discussion, including the ionic strength, protein effect, etc., must be based on the interphasic conditions. For example, most proteins will be present at the interphase in concentrations that far exceed their bulk value; the activity of water will be determined by the hydrophobicity of the interphase; ions will specifically adsorb; and so on. It is well known that upon covalent immobilization at the surface, the dissociation constant of the acid–base indicator changes by as much as $3 \, pK$ units. This shift clearly illustrates the dramatic effect that the interphase has on the ionization equilibria. It is perhaps the most serious problem with optical sensors that the surface concentration of any species is related to its corresponding bulk activity value through an adsorption isotherm which, with the exception of Henry's law, is a highly nonlinear relationship. It is also known[17] that the surface pH differs from the bulk value due to the electrostatic repulsion,

$$pH_{surf} = pH_{bulk} + Ne\phi/2.3 \, RT \qquad (5\text{-}37)$$

The surface potential ϕ depends in turn on the concentration profiles of all ionic species present at the interphase, i.e., diffuse-layer, specifically adsorbed, species, and on the surface ionizable groups. Thus the surface itself possesses acid–base properties that are reflected in the sensor response.

Can one eliminate the protein error by separating the sensor from the sample by a dialysis membrane? In fact, what is the effect of any membrane on the performance of a sensor for ionic species? We have seen earlier (Section 4.1.1.2) that exclusion of one type of ion (here, ionized protein) by a membrane generates the Donnan potential at that membrane [equation (4-21)]. This potential, in turn, affects the distribution of all ions on both sides of the membrane. Thus a potential difference of 20 mV, which is common on dialysis membranes at protein and electrolyte concentrations during, e.g., hemodyalysis, will lead to a pH difference of 0.33 units between the interior and the sample. The indicator will be protected from the undesirable effect of the proteins, but the existence of the Donnan potential at the boundary between the polymer matrix containing the indicator and the sample containing the excluded ion will lead to a considerable error of electrochemical origin, despite the fact that it is an optical sensor.

So far, we have ignored the last term in equation (5-35), the logarithm of the activity of water. This is legitimate in aqueous solutions in which the ionic strength does not exceed approximately 4 M. However, in mixed solvents or in concentrated electrolyte solutions this term becomes dominant. This phenomenon is common, to a varying degree, for all water–organic solvent mixtures and is due to the fact that the activity term of free, uncoordinated water decreases as the concentration of the organic solvent increases. Because solvation equilibria are interdependent for all species and for interphases present in the system, it is not surprising that the activities of all species are affected in a most profound way by the presence of organic solvents. The degree to which individual ions, including the indicators, are affected by the medium depends very much on their structure, which gives rise to a multiplicity of so-called acidity functions obtained with different types of indicator.[18]

In spite of these deficiencies optical sensing of ionic species has some attractions. The main one is that a determination can be conducted with a monolithic probe (i.e., without a reference electrode needed for an electrochemical measurement). If one accepts the lower accuracy of such measurements in all but very dilute aqueous solutions, it is possible to use optrodes for semiquantitative sensing of ions. In summary, it can be stated that experimentally determined pH errors due to the liquid junction are always smaller than those caused by the effect of ionic strength, organic solvent, or proteins on the optical indicators. This is the main reason why conventional measurements of pH and other ions by electrochemical methods are always preferred to colorimetric techniques. There is no reason to believe that such a preference should not apply also to optical sensors.

5.3.2. Gas Sensors

The number of possible combinations of different chemistries and different optical arrangements is obviously very large and no attempt is made here to present an exhaustive review. A few examples of optical gas sensors have been selected to point out some interesting features of sensors based on absorption, fluorescence, and evanescent wave principles.

5.3.2.1. Absorption Sensors

The most straightforward arrangement is the analog of the Severinghaus electrode (Section 4.1.5.1) or its conductimetric analog (Section 4.3.3) in which the internal pH-sensitive element is a pH optrode. It was made and tested for response to NH_3, but the extension to other acidobasic gases is obvious.[19] The internal electrolyte (Figure 5-19) contains p-nitrophenol as indicator ($pK_a = 7.8$). The p-nitroxyphenolate anion absorbs strongly at 404 nm. The dynamic range of this sensor can be adjusted by the concentration of the indicator, which acts as its own buffer. Thus for $[\text{In}] = 10^{-5}$ M it is 0.25–1.0 mM while for 10^{-2} M it lies between 0.005 and 0.1 mM. Like the Severinghaus electrode it can be used either in liquid or in gas samples with interferences coming only from other acidobasic gases. The response time is relatively long, about 4 min.

The composition of the solution in the inner compartment is constant, so the activity coefficient ratio and the activity of water in equation (5-35) is presumed to be constant and no problems due to the variation of these two terms are expected. It should be noted that the inner solution can be any solvent, including a totally nonconducting one. In that case, other

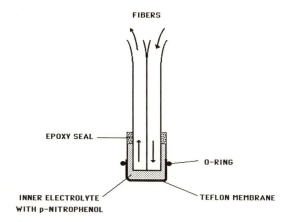

FIBERS

EPOXY SEAL —

O-RING

INNER ELECTROLYTE
WITH p-NITROPHENOL

TEFLON MEMBRANE

Figure 5-19. Optical gas sensor for CO_2 based on the Severinghaus principle. (After Ref. 19.)

gases which undergo complexation with a suitable indicator but no dissociation to ions could be sensed by this principle.

A layer of hemoglobin adsorbed on an aqueous gel of Sephadex has been used for sensing oxygen.[20] The optical system uses a bifurcated fiber and a reflector (Figure 5-16c). The measurement is carried out at two wavelengths, 435 nm (corresponding to deoxyhemoglobin) and 405 nm (corresponding to oxyhemoglobin). The two-wavelength measurement reduces the drift in the system due to fluctuations of the intensity of the source and due to mechanical effects on the optical alignment. A nonlinear response in the range 0–120 torr O_2 with 3 min response time is obtained. The biggest problem of this sensor is the short lifetime (2 days on the shelf at room temperature) due to the irreversible degradation of oxyhemoglobin to methemoglobin.

The sensor for polar solvent vapors gives rise to an interesting point.[21] It is based on a reversible protonation reaction between 3,3-diphenylphthalide(I) and a proton donor, here bis-phenol (BP),

$$3,3\text{-diphenylphthalide} + BP \Leftrightarrow \text{triphenylcarbenium}^+ + BP^-$$

Both the indicator and acid are incorporated in a thin PVC film cast on a microscope slide. Its absorbance was measured in a pocket-size photometer. Embodiment in any other optical system involving fibers and/or waveguides is quite clear.

The absorbance of triphenylcarbenium ion, which is intensively colored ($\lambda = 610$ nm), is used for the measurement. Again equation (5-35) applies, but this time we are dealing with reaction in (solid) organic solvent. Therefore, solvation effects [the fourth term in equation (5-35)] play a dominating role. Consequently, the response of the sensor is strongly affected by the relative humidity. The polar organic vapor molecules (i.e., alcohol or acetone) either compete for the solvation of the indicator molecule and change the fourth term or, less likely, affect the activity coefficient ratio (log f_A/f_{HA}) of the indicator. The device is not particularly sensitive: the dynamic range for ethanol and acetone is 2–10%. The response time is on the order of a few seconds.

5.3.2.2. Evanescent Wave Sensors

The optical waveguide sensor for NH_3, in which the indicator is applied as a neat solid film on the outer surface of a glass capillary tube

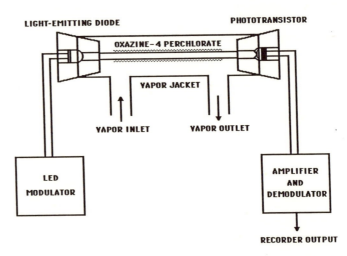

Oxazine-4 perchlorate

(Figure 5-20), is another example of an evanescent wave sensor.[22] We have not discussed a thin-wall (300-μm) glass capillary as a possible waveguide, but there is no reason why this geometry could not be used. The sensor also uses a blue (560-nm) light-emitting diode and a phototran-

Figure 5-20. Optical sensor for NH_3 using a capillary waveguide. The signal is obtained from attenuation of the evanescent component of light by selective coating. (After Ref. 22.)

sistor as the source and detector, respectively, which makes the whole system very compact. The equilibria which operate in the indicator film and on which this sensor is based again involve water,

$$NH_3 + H_2O = NH_4^+ + OH^-$$

$$NH_3 + HDye = NH_4^+ + Dye^- (610 \text{ nm})$$

As a result of the first reaction water vapor is a major interferant. Although it has not been reported, it is expected that this sensor would also be sensitive to other polar solvent vapors in the same way as the device described above.

A humidity sensor using an almost identical optical setup has been described.[23] The reaction is the well-known hydration equilibrium of cobalt(II) chloride salts,

$$CoCl_2 + 6H_2O \Leftrightarrow CoCl_2 \cdot 6H_2O$$

700 nm   500 nm

The salt is incorporated in poly(vinylpyrrolidone) film cast on the outer surface of the capillary (Figure 5-20). Since only one equilibrium constant is involved, the sensor shows a sigmoidal response between 85 and 80% relative humidity. The response time is tens of seconds.

5.3.2.3. Fluorescent Sensors

Fluorescence measurements have the advantage of a high signal-to-noise ratio owing to the frequency and geometrical shift of the emitted radiation with respect to the exciting light. The sensors described here are based on the quenching effect. Thus the determinand is a quencher Q_1 and the indicator is a fluor F which, when suitably excited, emits characteristic fluorescence. The reaction between the two is generally reversible,

$$F + Q_1 \xrightleftharpoons{K} FQ_1, \qquad K = [FQ_1]/[F][Q_1] \qquad (5\text{-}38)$$

There are no ionic reactions involved, which means that any changes in the activity coefficient ratio (for the two neutral forms of the fluor) would be small. We can therefore justifiably use concentrations instead of activities. The total fluor concentration C_F is given by

$$C_F = [F] + [FQ_1] \qquad (5\text{-}39)$$

Only the "free" indicator fluoresces, so equations (5-4), (5-38), and (5-39) yield the Stern–Volmer equation

$$(I_0 - I)/I = K_{SV}[Q_1] \tag{5-40}$$

in which the Stern–Volmer constant is defined as

$$K_{SV} = \gamma Q \varepsilon L K \tag{5-41}$$

It depends both on the quenching reaction and on the experimental parameters. If more quenchers are involved, equation (5-40) can be expanded in the form

$$(I_0 - I)/I = K_{SV1}[Q_1] + K_{SV2}[Q_2] + \cdots \tag{5-42}$$

By using at least as many sensors with different K_{SV} values as there are quenchers, one can obtain enough equations to determine the mixture. This approach has been used for the determination of mixtures of halothane (2-chloro-2-bromo-1,1,1-trifluoroethane) and oxygen (Figure 5-21).[24] This is an important problem for the management of anesthesia. Two sensors with different Stern–Volmer constants K_A and K_B have been constructed and their relative fluorescences [equation (5-42)] measured in pure quencher gas are α and β, respectively. By simple algebra the concentration of halothane [H] and oxygen [O] can be expressed as

$$[H] = (\alpha K_B^O - \beta K_A^O)/(K_A^H K_B^O - K_A^O K_B^H) \tag{5-43}$$

and

$$[O] = (\beta K_A^H - \alpha K_B^H)/(K_A^H K_B^O - K_A^O K_B^H) \tag{5-44}$$

Halothane is effectively blocked by a 12-μm-thick Teflon membrane. If

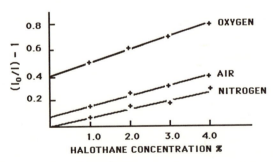

Figure 5-21. Response of a fluorescent gas sensor to halothane according to equation (5-42). (Reprinted from Ref. 24 with permission of the American Chemical Society.)

such a membrane is used to shield sensor B, then the above equations simplify to

$$[H] = (\alpha K_B^O - \beta K_A^O)/K_A^H K_B^O \qquad (5\text{-}45)$$

and

$$[O] = \beta/K_A^H \qquad (5\text{-}46)$$

When the Stern–Volmer constants for oxygen pertaining to both sensors are the same, equation (5-45) further simplifies to

$$[H] = (\alpha - \beta)/K_A^H \qquad (5\text{-}47)$$

Several different fluors have been tested for this application and the best results obtained for decacyclene ($\lambda_{ex} = 385$ nm; $\lambda_f = 510$ nm; $K_{SV} = 0.62$) imbedded in silicone rubber film. Fluoranthene, which has a higher value of K_{SV}, requires UV excitation ($\lambda_{ex} = 360$ nm) and is therefore less convenient to use. The response time is between 15 and 20 s for both gases and

Decacyclene

the precision is better than that obtained with the Clark electrode. A sterilizable oxygen probe based on this principle with performance characteristics equal to or better than an amperometric oxygen electrode has been constructed.[25]

5.3.3. Biosensors

The combination of selectivity of biochemical origin with high intrinsic selectivity of optical spectroscopy makes optical sensors potentially the most selective of all types of chemical sensors. Moreover, the coupling requirements between primary interactions in the selective layer and the transducer part of the sensor are relatively simple: the molecule which provides the signal (absorber) must merely be present in the optical path in order to be counted.

Any optical principle discussed in Sections 5.1 and 5.2 or their combination can be used in the actual implementation of an optical biosensor. It is sometimes difficult to distinguish between a spectrophotometric determination conducted in a conventional way in the cuvette and an identical assay in which light is guided around by an optical fiber or a waveguide. There are many reports in the literature in which the procedure or an instrument is labeled an "optical biosensor" only because a waveguide is used somewhere in the experimental setup. It really does not matter too much which label is attached to the procedure as long as it is done in a satisfactory way; however, from the standpoint of this book it is sometimes difficult to find the dividing line. We will definitely omit all "optical biosensors" in which one or more manipulations of the sample are done in order to obtain a data point. Such measurements are simply biochemical assays with an optical end-determination.

5.3.3.1. Optical Enzyme Sensors

By definition these are again steady-state sensors in which information about the concentration of the determinand is derived by measuring the steady-state value of a product or a substrate. The sensor itself does not consume or produce any of the species involved in the enzymatic reaction, so it is a *zero-flux boundary sensor* very much like a potentiometric enzyme sensor (Section 4.1.4) or an enzyme thermistor (Section 2.1). The optimum conditions for the operation of this sensor, particularly the thickness of the enzyme layer, could be found by again solving the set of diffusion-reaction equations for a given geometry [equation (1-43)]. There is, however, an important difference: while in the potentiometric sensor it is only the concentration at the surface of the sensor which generates the signal, in the case of an optical sensor it can be the surface and/or the bulk of the enzyme layer. Obviously, because the number of absorber molecules in the bulk is higher than at the surface, in principle the optical enzyme sensor can achieve a lower detection limit and higher sensitivity than the corresponding potentiometric sensor.

One example of an optical enzyme sensor[26] is a bifurcated optical fiber (Figure 5-22) which delivers and collects light to and from the site of the enzymatic reaction. The enzyme, alkaline phosphatase (AP), catalyzes hydrolysis of *p*-nitrophenyl phosphate to *p*-nitrophenoxide ion which is being detected ($\lambda = 404$ nm),

$$p\text{-nitrophenyl phosphate} + H_2O \Rightarrow p\text{-nitrophenoxide} + \text{phosphate}$$

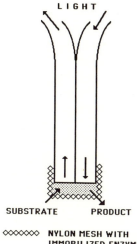

LIGHT

SUBSTRATE   PRODUCT

◇◇◇◇◇◇ **NYLON MESH WITH IMMOBILIZED ENZYME**

Figure 5-22. Schematic diagram of an enzyme optrode. The steady-state concentration of the optically active product is obtained at and inside the mesh, which acts also as a scatterer for the light. Another possibility is to confine the enzyme to the bulk of the space between the membrane and the end of the bundle. (After Ref. 26.)

The enzyme is immobilized on a nylon mesh, which acts also as a diffuse reflector for the light. The dynamic range of this sensor is between 10^{-5} and 4×10^{-3} M. Although the primary process determining the steady-state concentration of the *p*-nitrophenoxide ion is the diffusion-reaction mechanism, which is governed by the concentrations of all participating species, the *detection* of its concentration is again subject to the limitations of the optical sensing of ionic species (Section 5.3.1). From this point of view it is a much sounder strategy to base the detection on the concentration of electrically neutral species, which are affected by variations in their activity coefficients far less than are the ions. The major interferant in this sensor is again the buffer capacity.

There have been far fewer optical enzyme sensors reported in the literature than enzyme electrochemical sensors. Even now one can already say that the design of optical sensors mimics the design of their electrochemical counterparts. However, the wider choice of optical parameters and greater sensitivity seem to favor optical transduction over any other type.

5.3.3.2. Immunochemical Sensors

This is a relatively recent area of study but with one of the highest payoffs. There are several conceptual and review papers which present possible routes to immunochemical optical sensors and analyze the potential problems in their realization.[27,28] The proposed optical approaches are relatively straightforward. The interaction of light with the absorber (which can be an antibody, hapten, antigen, or second antibody) can take place either in the bulk or at the surface. Detection is invariably carried out in a fluorescent mode for reasons of higher sensitivity.

The main problem seems to be the immunological equilibrium itself (Section 1.2.3). Direct binding of the antigen to an immobilized high-affinity, polyclonal antibody results in rapid and irreversible saturation of the surface. The dissociation requires a change of solution and the dynamic range is narrow. For this reason several papers have described a system in which the initial rate of binding (adsorption) is taken as the measure of the antigen concentration. The antibody/antigen complex can be measured, for example, by TIRF (labeled or intrinsic) for which the quantitative relationships have already been developed.[29] The salient feature of this theory is that it allows one to differentiate between the contributions to the total fluorescence from the bound (N_{bound}) and free (soluble) (N_s) labeled protein,

$$\frac{N_{bound}}{N_s} = \frac{Q_{bound} \Gamma_p}{Q_s C_p \delta} [1 - \exp(-2\delta/d_p)] \tag{5-48}$$

where d_p is the depth of penetration of the evanescent wave [equation (5-19)], the Q quantities are the fluorescence quantum yields [equation (5-3)], Γ_p is the surface concentration of the bound species, and δ is the thickness of the adsorption layer. As long as the latter is small,

$$[1 - \exp(-2\delta/d_p)] = 2\delta/d_p \tag{5-49}$$

For UV excitation d_p is small and most of the signal originates from the bound protein.

Competitive binding realized within the sensor itself seems to be the only true reversible immunosensor reported to date. It employs the "model immunochemical reaction" between glucose and concanavalin A.[30] The operational scheme of this sensor is shown in Figure 5-23. The immunochemical reaction takes place inside a cavity placed at the end of the bifurcated fiber. The walls of the cavity are formed by a dialysis membrane, which allows passage of low-molecular-weight haptens but cuts off haptens attached to high-molecular-weight polymers. The fiber has a low numerical aperture (narrow cone of acceptance), so that the fluorescence is excited only by molecules which can move freely inside this cavity and in the optical path but not by molecules attached to the wall. This

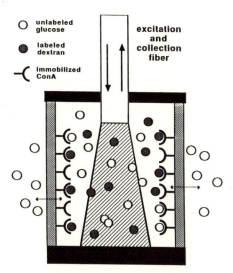

Figure 5-23. Immunochemical sensor for glucose based on competitive binding by concanavalin A of *freely permeable* glucose and a *confined* labeled dextran. (After Ref. 30.)

fluorescence is coupled back to the fiber and quantified in a normal manner. There are two schemes that can be employed. In the first the antibody is attached covalently to the inside surface of the membrane. The fluorescently labeled hapten is attached to the high-molecular-weight polymer, which confines it only to the interior of the cavity. On the other hand, low-molecular-weight hapten (the determinand) can diffuse freely through the membrane and compete with the labeled hapten for the available antibody binding sites. Since only the mobile hapten can be detected by the incoming light, the fluorescent intensity reflects the competition equilibrium.

In the second scheme the hapten is immobilized at the inner surface. The fluorescently labeled antibody, which is again trapped inside, is bound either to the immobilized hapten or to a freely diffusing hapten that enters the cavity from the outside. In both cases the increase in the exterior concentration of hapten leads to the increased intensity of fluorescence detected by the fiber.

5.3.4. Electrooptical and Optomechanical Sensors

In this section we examine briefly several phenomena which have shown promise for further development as chemical sensors. They involve light, but the transduction principle differs somewhat from those described in Sections 5.1 and 5.2.

5.3.4.1. Photoluminescent Schottky Diodes

When a semiconductor is illuminated with light of sufficient energy to excite electrons from the valence band to the conduction band, a hole–electron pair is generated. The relaxation can take several pathways, some of which are nonradiative. Radiative transition leads to *photoluminiscence* (Figure 5-24). This is an analogous situation to the interaction of electromagnetic radiation with molecules which leads to fluorescence (Section 5.1). Nonradiative recombination and radiative transition are clearly two competing processes.

A Schottky barrier junction is constructed, for example, by depositing Pd on n-CdS. Its simplified energy-band diagram is shown in Figure 5-25 together with a voltage source, which applies a suitable bias across the diode. As we have described repeatedly, hydrogen dissolves reversibly in Pd and in so doing changes its work function. Section 4.1.5.2 treated the semiconductor/(insulator)/metal junction under conditions of zero electric field in the insulator, in the so-called "flat band condition." On the other hand, the Schottky junction (metal/semiconductor) used in photo-luminiscent studies is under an electrical bias, which means that the energy bands are bent and a region of non-zero electric field (space charge) is formed in the semiconductor (Appendix C). Its presence is important for the effect to be used in chemical sensing.[31] The hole–electron pair within this region is separated by the electric field, thus reducing the probability of radiative transition in the recombination process. The space charge is nonemissive, and is therefore called a "dead layer" in the context of this

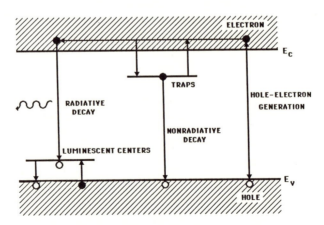

Figure 5-24. Excitation of a hole–electron pair in a wide-bandgap semiconductor. Two relaxation processes dominate: nonradiative recombination and radiative transition leading to luminiscence. Note the analogy with the absorption/fluorescence process occurring at the molecular level (Figure 5-4).

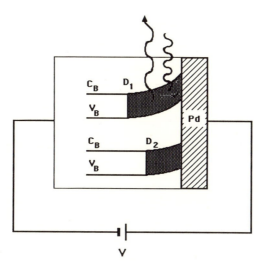

Figure 5-25. Modulation of photoluminiscence from the Pd/n-CdS Schottky diode. The biasing voltage V sets the baseline emission. The two thicknesses of the "dead layer," D_1 and D_2, correspond to two levels of the work function of Pd. (After Ref. 25.)

discussion. In other words, the relative intensity of the emitted light is decreased as the thickness of the space charge increases, and *vice versa*. The semiempirical relationship describing this effect is formulated in terms of fluorescent quantum yields Q_{air} and Q_{H_2} for the illuminated junction in air and in hydrogen, respectively,

$$Q_{air}/Q_{H_2} = \exp[-(\alpha + \beta)\,\Delta D] \tag{5-50}$$

where α and β are the absorptivities of the semiconductor for the exciting and emitting radiations while ΔD is the change in the dead-layer thickness. For a constant applied bias the thickness of the dead-layer region is given by equation (4-86). It is modulated by the change in the work function of Pd (Section 4.1.5.2).

This concept can be extended to other semiconductors[32] in which the photoluminiscent "dead layer" is modulated in a more subtle way. For example, n-GaAs derivatized with ferrocene responds to volatile oxidants (such as halogens) according to the above mechanism, because the ferrocene molecule facilitates partial transfer of an electron from the semiconductor to the adsorbed molecule. This separation of charge creates an additional electric field, which affects the dead-layer thickness at the surface and the intensity of photoemission from this layer [equation (5-50)]. In principle, any gas with Lewis acid/base characteristics should yield a similar response. It is interesting to note the analogy of this mechanism

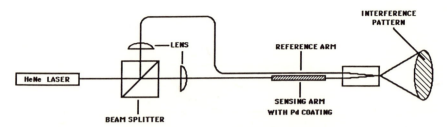

Figure 5-26. Optomechanical interference sensor according to Ref. 33.

with that on which semiconducting oxide sensors (Section 4.3.2) are based. There, modulation of the surface concentration of the adsorbed oxygen affects the thickness of the space-charge region, and hence the conductivity.

5.3.4.2. Optomechanical Sensor

This concept[33] employs the interferometric principle, in which two beams of coherent light are combined to form a constructive interference pattern (Figure 5-26). The two beams pass through two fibers, one coated with a thin (10-μm) layer of Pd approximately 3 cm long while the other serves as reference. The interaction between palladium and hydrogen has been used extensively throughout this book because it is a very versatile model, particularly for the effects based on changes in the work function. In this case, however, we shall examine mechanical changes taking place when hydrogen is dissolved in palladium. This causes a relative change in the lattice constant a_0 (atom spacing) proportional to the amount of palladium hydride C_{PdH} and therefore to the partial pressure of hydrogen,

$$\Delta a_0/a_0 = 0.026 \, \Delta C_{PdH} \tag{5-51}$$

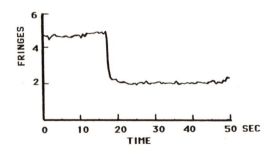

Figure 5-27. Response of an optomechanical sensor to a concentration step of hydrogen from 10 to 20 ppb. Nitrogen was used as carrier gas. (Reprinted from Ref. 33 with permission of the American Institute of Physics.)

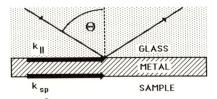

Figure 5-28. Origin and use of the plasmon resonance effect. The oscillating plasmon k_{sp} is induced by resonant phase matching with the parallel electrical vector of the reflected light $k_{||}$.

The macroscopic effect of a change in the lattice constant on a layer of Pd deposited on an optical fiber is the appearance of both radial (U) and axial (W) stress in that layer. This stress is transmitted to the fiber stretching it, and thus increasing the optical path ΔL,

$$\Delta L/L = W - 0.5\, n^2 [\, U(P_{11} - P_{44}) + W(P_{11} - 2P_{44})] \qquad (5\text{-}52)$$

This increase is observed as a shift in the interference lines. Constants P are Pockel's coefficients. A realistic estimate of the dynamic range of this sensor is <1 ppm to 10^4 ppm hydrogen (Figure 5-27).

It would appear that another type of hydrogen sensor is the last thing needed. However, both optomechanical and photoluminiscent concepts are important in the sense that they may lead to the development of other types of similar devices for different gases. It is conceivable that other gas-sensitive coatings may exhibit similar properties.

5.3.4.3. Plasmon Resonance

When a thin (500-Å) film of metal is irradiated with coherent light which is TM polarized (magnetic field oriented perpendicularly to the plane defined by the surface normal and the direction of propagation), oscillating charges may be induced at its surface (Figure 5-28). These are

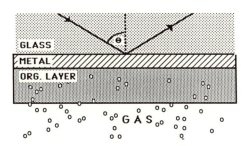

Figure 5-29. Plasmon resonance used for the detection of gases. A silicon–glycol copolymer has been used to detect halothane in the concentration range 100–1000 ppm.

called non-radiative surface plasmons,[34] which move along the surface but do not radiate energy. Their intensity decays exponentially in the direction perpendicular to the surface. In that respect they are similar to evanescent waves. Surface plasmons exist for frequencies at which the real part of the dielectric constant of the metal ε_m is less than the dielectric constant of the neighboring dielectric ε_i (here glass). The relationship between the frequency and the value of the propagating vector k_1 for the plasmon at the glass–metal surface is (c is velocity of light)

$$\omega = ck_1(1/\varepsilon_m + 1/\varepsilon_i)^{1/2} \tag{5-53}$$

For light striking the surface at an incident angle Θ, the wave vector of the light which is parallel to the surface is given by

$$k_{||} = \omega c^{-1} \varepsilon_{air}^{1/2} \sin \Theta \tag{5-54}$$

Vector $k_{||}$ couples the energy of the photon to the propagating surface plasmon.

For a very thin metal, the dielectric constants of the two dielectric media affecting the propagation of the plasmon are different. Velocities k_1 and $k_{||}$ can be matched by varying the angle of incidence [equation (5-54)] and results in plasmon resonance. It appears as the minimum on a plot of the intensity of the reflected light as a function of the angle of incidence. The value of the dielectric constant at the air–metal interface thus determines the shift of the resonance.

This phenomenon has been proposed and tested as a mechanism for the detection of gases (Figure 5-29) and of adsorbed proteins (Figure 5-30).[35]

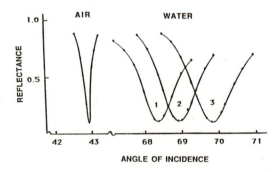

Figure 5-30. The plasmon resonance effect applied to absorption of proteins. Resonance curves were obtained on a 600-Å film of silver on glass: (1) clean surface, (2) shift due to irreversible adsorption of 50 Å (estimated) thick layer of human IgG, (3) further shift obtained upon addition of anti-human IgG. (Reprinted from Ref. 35 with permission of Elsevier.)

Glossary for Chapter 5

A	Absorbance
a_0	Lattice constant
c	Velocity of light
d_p	Depth of penetration (of evanescent wave)
E	Intensity of electric field
E_λ	Photon energy
h	Planck's constant
I_F	Fluorescent light intensity
k	Absorption index
k_1	Surface plasmon vector
K_{SV}	Stern–Volmer constant
M	Number of (fiber) modes
n	Refractive index
NA	Numerical aperture
P_{11}	Pockel's coefficient
Q	Quantum efficiency
r	Radius
U	Radial stress
W	Axial stress
δ	Thickness of adsorption layer
ε	Molar absorptivity
ε_i	Dielectric constant of insulator
ε_m	Dielectric constant of metal
γ	Collection factor
λ	Wavelength
ν	Frequency
Θ_1	Angle of incidence
Θ_c	Critical angle

References for Chapter 5

1. J. R. Lakowicz, *Principles of Fluorescence Spectroscopy*, Plenum Press, New York, 1983.
2. A. T. Tu, *Raman Spectroscopy in Biology*, Wiley–Interscience, New York, 1982.
3. M. Born and E. Wolf, *Principles of Optics*, Pergamon Press, New York, 1980.
4. A. W. Snyder and J. D. Lowe, *Optical Waveguide Theory*, Chapman and Hall, New York, 1983.
5. L. Rouchi and F. M. Schaggi, *Adv. Electron. Electron Phys.* 51 (1980) 63.
6. P. Kubelka and F. Munk, *Z. Tech. Phys.* 12 (1931) 593.
7. H. G. Hecht, *Appl. Spectrosc.* 37 (1983) 348.
8. W. R. Seitz, *Anal. Chem.* 56 (1984) 16A.
9. O. S. Wolfbeis, *Fresenius Z. Anal. Chem.* 325 (1986) 387.
10. K. Newby, W. M. Reichert, J. D. Andrade, and R. E. Benner, *Appl. Opt.* 23 (1984) 1812.

11. N. J. Harrick, *Internal Reflection Spectroscopy*, Harrick Sci. Corp., New York, 1979.
12. Y. L. Cheng, B. K. Lok, and C. R. Robertson, in: *Surface and Interfacial Aspects of Biomedical Polymers* (J. D. Andrade, ed.), Vol. 2, Plenum Press, New York, 1985.
13. V. Hlady, R. A. Van Wagenen, and J. D. Andrade, in: *Surface and Interfacial Aspects of Biomedical Polymers* (J. D. Andrade, ed.), Vol. 2, Plenum Press, New York, 1985.
14. O. S. Wolffbeis and H. Offenbacher, *Sensors and Actuators* 9 (1986) 85.
15. J. Janata, *Anal. Chem.* 59 (1987) 1351.
16. N. Opitz and D. W. Lübbers, *Sensors and Actuators* 4 (1983) 473.
17. J. T. Davies and E. K. Rideal, *Interfacial Phenomena*, Chapter 2, Academic Press, Orlando, 1963.
18. R. H. Boyd, in: *Solute–Solvent Interactions*, Vol. 1 (J. F. Coetzee and C. D. Ritchie, eds.), Dekker, New York, 1969.
19. M. A. Arnold and T. J. Ostler, *Anal. Chem.* 58 (1986) 1137.
20. Z. Zhujun and W. R. Seitz, *Anal. Chem.* 58 (1986) 220.
21. F. L. Dickert, E. H. Lehman, S. K. Schreiner, H. Kimmel, and G. R. Mages, *Anal. Chem.* 60 (1988) 1377.
22. J. F. Giuliani, H. Wohltjen, and N. L. Jarvis, *Opt. Lett.* 8 (1983) 54.
23. D. S. Ballantine and H. Wohltjen, *Anal. Chem.* 58 (1986) 2883.
24. O. S. Wolfbeis, H. E. Posch, and H. W. Kroneis, *Anal. Chem.* 57 (1985) 2556.
25. H. W. Kroneis and H. J. Marsoner, *Sensors and Actuators* 4 (1983) 587.
26. M. A. Arnold, *Anal. Chem.* 57 (1985) 565.
27. J. D. Andrade, R. A. Van Wagenen, D. E. Gregonis, K. Newby, and J.-N. Lin, *IEEE Trans. Electron Devices* ED-32 (1985) 1175.
28. J. F. Place, R. M. Sutherland, and C. Dähne, *Biosensors* 1 (1985) 321.
29. V. Hlady, D. R. Reinecke, and J. D. Andrade, *J. Colloid Interface Sci.* 111 (1986) 555.
30. J. S. Schultz, S. Mansouri, and I. J. Goldstein, *Diabetes Care* 5 (1982) 245.
31. M. K. Carpenter, H. Van Ryswyk, and A. B. Ellis, *Langmuir* 1 (1985) 605.
32. H. Van Ryswyk and A. B. Ellis, *J. Am. Chem. Soc.* 108 (1986) 2454.
33. M. A. Butler, *Appl. Phys. Lett.* 45 (1984) 1007.
34. J. D. Swalen, J. G. Gordon, II, M. R. Philpott, A. Brillante, I. Pockrand, and R. Santo, *Am. J. Phys.* 48 (1980) 669.
35. B. Lindberg, C. Nylander, and I. Lundstrom, *Sensors and Actuators* 4 (1983) 299.

Appendixes

A

Survey of Thermodynamics[1-3]

The thermodynamical laws are statistical laws, which means that they describe large assemblies of particles. A *system* is defined as an arbitrary part of the Universe with defined boundaries. If neither heat nor matter is exchanged between the system and its surroundings, it is called an *isolated system*. If matter cannot cross its boundaries it is said to be *closed*, and *vice versa*. If it is thermally insulated, it is an *adiabatic system*.

Any system can be described uniquely by a minimum number of parameters, called *state variables*. Their relationships are called *equations of state*.

For example, an ideal gas inside a piston compartment is a system. It is described uniquely by the equation of state for ideal gas

$$pV = RT \tag{A-1}$$

where the state variables are p (pressure), V (volume), and T (temperature); R is the universal gas constant.

A.1. The First Law of Thermodynamics

Each system contains a certain amount of internal energy U which resides in the kinetic energy of the individual molecules, in the energy of their bonds, etc. The first law of thermodynamics defines the relationship between the *work* W done by the system ($W > 0$) and the *heat* Q absorbed by the system ($Q > 0$). We note that the sign of these two variables relates to the exchange of these two quantities with the environment of the system. The first law simply states that the change in the internal energy of a

287

system dU equals the difference between the work done and the heat received by the system,

$$dU = dQ - dW \tag{A-2}$$

For an adiabatic system $dQ = 0$ and $dU = -dW$. If the only type of work which a system can do is mechanical work, then

$$dU = dQ - P \, dV \tag{A-3}$$

Other examples of work W_{other} might be *electrical* work dW_e, associated with moving charge q across a potential difference $d\phi$,

$$dW_e = q \, d\phi \tag{A-4}$$

or work dW_s associated with an increase in surface area dA against surface tension γ,

$$dW_s = \gamma \, dA \tag{A-5}$$

Most chemical reactions take place at constant pressure. In that case the change of internal energy is related to a new thermodynamical function called *enthalpy H*,

$$H = U + PV + W_{other} \tag{A-6}$$

It has a physical meaning of internal energy *when the heat is absorbed by the system under constant pressure,*

$$dH = dQ_P \tag{A-7}$$

When heat is supplied to a system it is stored in the kinetic energy of its molecules and the *temperature T* changes,

$$dT = dQ/C \tag{A-8}$$

The actual change of temperature depends on the heat capacity C, which is again dependent on whether the process is conducted under constant pressure (C_P) or at constant volume (C_V). Therefore

$$C_P = dQ_P/dT = (\partial H/\partial T)_P \tag{A-9}$$

Both quantities C_P and C_V are constant over only a narrow range of temperatures.

In summary, the first law of thermodynamics is an empirical accounting statement which relates the internal energy of a system to the heat supplied to it and the work done by it, regardless of the path by which this transaction has been accomplished.

A.2. The Second Law of Thermodynamics

This law establishes the directionality of the processes taking place in the system by defining a new function, *entropy S*,

$$dS = dQ_{rev}/T \tag{A-10}$$

where dQ_{rev} is the reversible change of heat. When a spontaneous (irreversible) process takes place the entropy of the system increases,

$$\Delta S \geqslant 0 \tag{A-11}$$

A more intuitive definition of entropy is in terms of probability: *a more random system has higher probability and higher entropy*,

$$S = k \ln W \tag{A-12}$$

where k is the Boltzmann constant and W is the measure of randomness of the system.

The *Gibbs free* (useful) *energy* is defined as

$$G = H - TS \tag{A-13}$$

For any process to take place *spontaneously* the Gibbs free energy must decrease,

$$\Delta G < 0 \tag{A-14}$$

Thus, the aforementioned directionality arises from the combination of equations (A-11) and (A-13). All closed systems tend to the state of maximum entropy (minimum free energy)—*equilibrium*.

A.3. Equilibrium

The state of a system in which there is no change of Gibbs free energy $(dG = 0)$ and entropy $(dS = 0)$ is called *equilibrium*. For an isothermal process equation (A-13) can be restated as

$$\Delta G = \Delta H - T \Delta S \tag{A-15}$$

If $T \Delta S > \Delta H$ the spontaneous process is said to be *entropy controlled*, and *vice versa*.

The free energy of an *ideal solution* of species x is given by

$$G_x = G_x^\circ + nRT \ln c_x \qquad \text{(A-16)}$$

Thus, the *standard state* G° is defined for unit concentration of x. For a simple chemical equilibrium

$$A \Leftrightarrow B$$

the *equilibrium constant K* is defined as

$$K = c_B/c_A \qquad \text{(A-17)}$$

It follows from equations (A-16) and (A-17) that

$$\Delta G^\circ = -RT \ln K \qquad \text{(A-18)}$$

where ΔG° is the *standard free-energy change for the reaction*.

A.4. The Chemical Potential

The free energy of an open system depends on temperature, pressure, and its *composition*. If the concentration of component i (number of moles n_i) in a system changes, the free energy of the system changes,

$$\mu_i = (\partial G / \partial n_i)_{T, P, j \neq i} \qquad \text{(A-19)}$$

This partial molar change of free energy is called the *chemical potential of species i*. If this species is charged (i.e., an ion) it is called an *electrochemical potential of species i*, because the change in the free energy of the system includes a component of *electrical work*. The electrochemical potential of an electron in a (solid) phase μ_e is equal to the Fermi potential of that phase. Because at equilibrium $dG = 0$,

$$dG = \sum \mu_i \, dn_i = 0 \qquad \text{(A-20)}$$

This is called the *Gibbs equation*, which is particularly important for understanding phase equilibria. A related expression, the *Gibbs–Duhem equation*, states that at equilibrium the change in the chemical potential of one component leads to a change in the chemical potentials of all the other components,

$$\sum n_i \, d\mu_i = 0 \qquad \text{(A-21)}$$

The concentration of components in a system can be expressed in different units and results in the definition of different standard states [equation (A-16)]. In a gas phase it is customary to introduce the *partial pressure* p_i,

$$p_i = P_i \Big/ \sum P_j \qquad \text{(A-22)}$$

The most general definition is the mole fraction X_i,

$$X_i = n_i \Big/ \sum n_j \qquad \text{(A-23)}$$

Molarity is the number of moles per one liter of solution, and *molality* is the number of moles per 1000 grams of solvent.

A.5. Ideal–Real Solutions

Up to this point only *ideal solutions* have been considered. If the solute does not behave as an inert particle (i.e., if it interacts with other solutes, including its own kind), then the energy of this interaction must be included in the internal energy of the system and equation (A-16) and those which follow must be defined in terms of *activities*,

$$a_i = f_i c_i \qquad \text{(A-24)}$$

The activity coefficient f_i expresses the extent of the nonideal behavior. It is defined as

$$\lim_{c \to 0} f = 1 \qquad \text{(A-25)}$$

These interactions exist in *real* systems for which the activity has the physical meaning of the *effective* concentration. Thus, $a_i = c_i$ *only* for dilute real solutions. The activity coefficient is always less than or equal to unity and assumes different values in different concentration scales.

A.6. Adsorption and Absorption

When a system contains two (or more) phases, the components of the system interact with these phases in one of two principal ways. The first possibility is that the component exists in both phases, in which case (at equilibrium) *the chemical potential of that component must be identical in both phases*. This follows from the Gibbs equation (A-20). The number of moles transferred from phase 1 equals the number of moles received by phase 2 ($dn_1 = -dn_2$); therefore

$$\mu_1 = \mu_2 \qquad (A\text{-}26)$$

This is the law governing all partitioning equilibria, including the electrochemical one in which the component is *absorbed* in a phase. Equation (A-26) is valid for all components, including the solvent. The chemical potential of the solvent is decreased by the presence of solute, so the transfer of the solvent across the membrane in the direction of the solvent chemical potential gradient takes place (flow is from the dilute solution to the concentrated solution) with concurrent increase in the pressure inside the more concentrated solution compartment. Pressure is one of the variables [see equations (A-6) and (A-15)] of the free energy, therefore it contributes to the new equilibrium state of the system. This pressure is called *osmotic pressure*,

$$\pi = RTC \qquad (A\text{-}27)$$

where C is expressed in mole liter^{-1}. It is a colligative property, meaning that it depends on the number of solute molecules. It is also additive, thus the final osmotic pressure is the sum of the partial osmotic pressures,

$$\pi = RT \sum_i C_i \qquad (A\text{-}28)$$

The second possibility is that the interface between the two phases presents an impervious barrier to the component (or components), in which case that component *adsorbs* (accumulates) at that interface. The amount of adsorbed species depends in a complex way on the interaction energies between the adsorbate and the interface and between the adsorbing species and all the other species competing for the interfacial area. The general relationship is

$$\beta_i c_i = \Re(\Gamma_i) \qquad (A\text{-}29)$$

Table A-1. Two-Dimensional Equations of State and Corresponding Adsorption Isotherms[a]

Isotherm	Equation of state	Surface pressure	Surface concentration
Henry's law	$p = RT\Gamma^i$	$Ka_X = p$	$Ka_X = RT\Gamma^i$
Langmuir	$p = -RT\Gamma_s \ln(1 - \Gamma^i/\Gamma_s)$	$Ka_X = \exp(p/RT\Gamma_s) - 1$	$Ka_X = \phi/(1 - \phi)$
Temkin	$p = g(\Gamma^i)^2$	$Ka_X = \exp[(4gp)^{0.5}/RT]$	$Ka_X = \exp(2g\Gamma^i/RT)$

[a] $\phi = \Gamma^i/\Gamma_s$, where Γ_s is the saturation value of Γ^i and g is the van der Waals repulsion.

where $\Re$ is the functional form of the adsorption isotherm, Γ_i the relative surface excess (surface concentration normalized with respect to an arbitrarily chosen species), and β_i is the interaction energy term containing the adsorption free energy G_{ads},

$$\beta_i = \exp(-G_{ads}/RT) \tag{A-30}$$

These equations relate the *bulk* concentration c_i in the sample to the surface concentration. Examples of some adsorption isotherms are given in Table A-1.

References for Appendix A

1. W. J. Moore, *Physical Chemistry*, Prentice-Hall, New York, 1966.
2. K. E. Van Holde, *Physical Biochemistry*, 2nd ed., Prentice-Hall, New York, 1985.
3. D. Freifelder, *Physical Chemistry*, Jones and Bartlett, Boston, 1982.

Survey of Reaction Kinetics

B.1. Equilibrium and Rate Equations

The reaction $A \Leftrightarrow B$ can be thought of as being composed of two processes, forward and reverse, which take place simultaneously with velocity v,

$$v_f = -dc_A/dt = k_f c_A \qquad \text{and} \qquad v_r = -dc_B/dt = k_r c_B \qquad \text{(B-1)}$$

The velocity equations are always expressed in concentrations rather than activities, because the time change of the activity coefficient is small. At equilibrium the two velocities are equal, therefore no *net* chemical change takes place and [cf. equation (A-17)]

$$k_f/k_r = c_B/c_A = K \qquad \text{(B-2)}$$

Thus, *time* is introduced as a new variable into the definition of chemical processes, including equilibrium ones. Although the absolute value of K may be the same, we can talk about *fast* or *slow* equilibria, meaning that the *rate constants* k_f and k_r are either high or low. Reactions involving two or more species are reactions of second and higher *order*. Hence for a second-order reaction $A + B \rightarrow C$ the rate law is

$$-dc_A/dt = -dc_B/dt = dc_C/dt = kc_A c_B \qquad \text{(B-3)}$$

For a first-order reaction the dimensions of the rate constant k are s^{-1}, while for a second-order reaction the dimensions are liter $mol^{-1} s^{-1}$. Second- (and higher-) order reactions can be converted to first-order reactions by having all components but one in a large molar excess. Those are called reactions of *pseudofirst order*. For a first-order reaction the time required to convert exactly one half of the original concentration is called a *half-time*, $t_{1/2}$, and is uniquely related to the first-order rate constant,

$$t_{1/2} = (\ln 2)/k \tag{B-4}$$

Any reaction scheme can be described by a set of first- or higher-order differential equations. These can be solved exactly for only relatively simple schemes. For more complicated sets it is necessary to use either analog computers or digital numerical methods.

B.2. Activation Energy

The rate of reaction depends on the temperature and on the value of the *activation energy* E_a,

$$k = A \exp(-E_a/RT) \tag{B-5}$$

The value of the activation energy determines the magnitude of the rate constant. It is approximately related to the enthalpy of formation of the *activation complex*, an intermediate in any transformation. Thus, reaction between A and B to form C can be expressed formally as progressing through the activation complex [ABC*],

$$A + B \rightarrow [ABC^*] \rightarrow C \tag{B-6}$$

For this reaction the activation energy in equation (B-5) can be expanded to the corresponding activation entropy ΔS^* and activation enthalphy ΔH^*,

$$k = [A' \exp(\Delta S^*/RT)] \exp(-\Delta H^*/RT) \tag{B-7}$$

The first exponential term in this equation is related to the frequency with which the components of the activation complex assume the correct orientation. It describes, among other things, the stereospecific fit of the components A, B, and C.

C

Survey of Solid-State Physics[1,2]

Semiconductors are the base materials of most modern chemical sensors. This appendix surveys the fundamental properties of semiconductors both in isolation and in combination with other materials.

When atoms are squeezed together to form a crystal (condensed state), their vacant and filled orbitals become so closely spaced that they merge to form the *conduction* and *valence* energy bands (Figure C-1). The energy difference between the occupied and filled orbitals is affected by this close packing. The energy difference between the bottom of the conduction band and the top of the valence band is called the *bandgap energy* E_g,

$$E_g = E_c - E_v \qquad \text{(C-1)}$$

The magnitude of E_g classifies the materials on the basis of their ability to conduct electricity. For $E_g \ll kT$ the material is a good conductor, and E_v and E_c overlap. Insulators, on the other hand, have a large value of E_g, typically in excess of 5 eV. Packets of coherent acoustic energy, called phonons, move through the crystal lattice and cause dissociation of electrons from the valence band to the conduction band leaving behind holes. In an *intrinsic semiconductor* the concentration of electrons n and holes p is given by

$$n = N_c \exp(-E_g/2kT) \qquad \text{and} \qquad p = N_v \exp(-E_g/2kT) \qquad \text{(C-2)}$$

where N_c and N_v represent the effective density of energy states in the conduction and valence bands, respectively. The Fermi level E_F, which expresses the average energy position of an electron in the material (Figure C-2), is defined as

$$E_F = (E_c + E_v)/2 + (kT/2) \ln N_v/N_c \qquad \text{(C-3)}$$

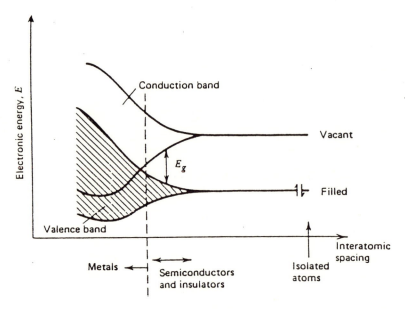

Figure C-1. Origin of the energy bands in solids explained by merging of the orbitals of isolated atoms. (Reprinted from Ref. 6, p. 630 with permission of Wiley.)

This latter level lies approximately in the middle of the bandgap, because $N_v \sim N_c$. For an intrinsic semiconductor the concentration of holes and electrons is identical, because they are generated by the same process. Therefore

$$np = n_i^2 = N_c N_v \exp(-E_g/kT) \qquad \text{(C-4)}$$

The analogy between self-dissociation of water to H^+ and OH^- is obvious. Addition of an "impurity" (or dopant) can also be regarded in the same light as the addition of weak base or a weak acid to water. The same condition of electroneutrality must apply,

$$N_A + n = N_D + p \qquad \text{(C-5)}$$

where N_A are negatively charged acceptor ions and N_D positively charged donor ions. The excess of holes or electrons then determines the *polarity* of the semiconductor (Figure C-2) as p-type or n-type. The position of the Fermi level is affected by the presence of impurities. It lies close to the conduction band for the n-type semiconductor and close to the valence band for the p-type semiconductor,

$$E_c - E_F = kt \ln N_c/N_D \qquad \text{and} \qquad E_F - E_v = kt \ln N_v/N_A \qquad \text{(C-6)}$$

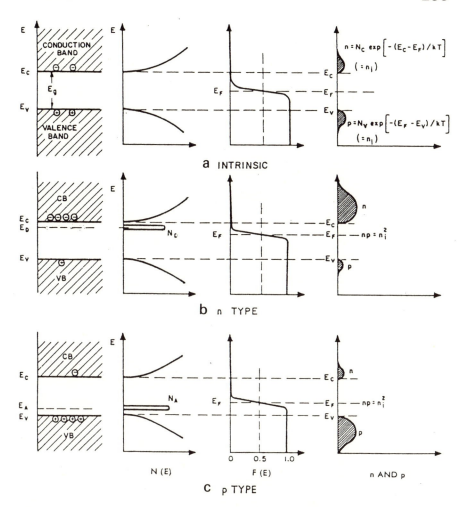

Figure C-2. Schematic diagram showing energy bands, density of states, Fermi–Dirac distribution, and carrier concentrations, for (a) intrinsic, (b) n-type, and (c) p-type semiconductors at thermal equilibrium. (Reprinted from Ref. 4, p. 33 with permission of Wiley.)

where N_D and N_A are the concentrations of the donor and acceptor impurities, respectively. The energy required to remove an electron from the Fermi level and place it in vacuum just outside the reach of the image forces is called the *work function*. It is a material constant defined by

$$E_F = \mu_e - e\phi = \mu_e - e\chi - e\psi \tag{C-7}$$

where ϕ is the inner electrostatic potential of the bulk, χ the dipole surface potential, and ψ is the outer electrostatic potential. For $\psi = 0$ the work function consists of the inner μ_e term and the surface $e\chi$ term. The Fermi level is often called the electrochemical potential of the electron [cf. Appendix A, equation (A-19)].

Transport of electricity in a semiconductor is carried out by both the holes and electrons, and the material conductivity is defined by

$$\sigma = q\mu_p[p] + q\mu_n[n] \qquad \text{(C-8)}$$

where the mobility of holes μ_p and electrons μ_n depends on the effective mass m^* of the carriers and is given approximately as

$$\mu \sim (m^*)^{-3/2} T^{1/2} \qquad \text{(C-9)}$$

The relationship to the carrier diffusion coefficient is

$$D_n = (kT/q)\,\mu_n \qquad \text{and} \qquad D_p = (kT/q)\,\mu_p \qquad \text{(C-10)}$$

Semiconductors are rarely used as such. When they are interfaced with other materials, junctions are formed. A junction between two semiconductors of opposite polarity is called a p–n junction. If the concentration of one type of dopant is much higher than that of the other (e.g., $N_A \gg N_D$), it is called an *abrupt junction* (Figure C-3). For approximately equal doping levels, we talk about a *two-sided junction*. A gradient of dopants is found in *graded junctions*. The concentrations of carriers on the two sides of a junction (subscripted) are

$$p_n = p_p \exp(qV_B/kT) \qquad \text{and} \qquad n_p = n_n \exp(-qV_B/kT) \quad \text{(C-11)}$$

where V_B is called the "built in" (contact) potential (Figure C-3). When an external potential is applied, two situations may develop: if the polarity of the applied potential follows the junction (i.e., positive potential to p-type and negative to n-type), then the junction is said to be *forward-biased* and a current will flow,

$$J_n = -qn\mu_n(d\phi/dx) \qquad \text{and} \qquad J_p = qp\mu_p(d\phi/dx) \qquad \text{(C-12)}$$

For a *reverse-biased junction* a depletion layer will form with width given by

$$W = (2\varepsilon_s V_B/qN_D)^{1/2} \qquad \text{(C-13)}$$

where N_D is the concentration of the donor. The capacity of this region is

$$C = dQ/dV = \left(\frac{q\varepsilon_s N_D}{2(V_B - V - kT/q)}\right)^{1/2} \qquad \text{(C-14)}$$

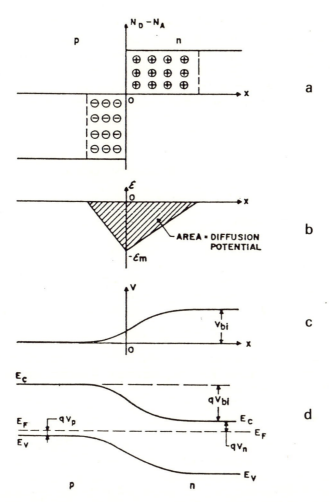

Figure C-3. Diagram of an abrupt p–n junction in thermal equilibrium: (a) impurity distribution, (b) electric field, (c) potential profile, (d) energy-band diagram. (Reprinted from Ref. 4, p. 86 with permission of Wiley.)

The Mott–Schottky relationship is used to evaluate the doping concentration from the plot of the inverse square of C vs. the applied voltage,

$$1/C^2 = \frac{2(V_B - V - kT/q)}{q\varepsilon_s N_D} \qquad (C\text{-}15)$$

The conditions at the metal–insulator–semiconductor (MIS) junction are discussed in Section 4 in greater detail. Here we examine only the condition which leads to *band bending*. It occurs when an external voltage is applied at such a junction and leads to the existence of a residual electric field in the semiconductor. This causes a redistribution of charge and the formation of a *space-charge region* near the semiconductor surface. The difference between this space charge and that existing at the n–p junction is that it does not depend on the polarity, because the insulator blocks the passage of current. The surface concentration of electrons and holes is again given, respectively, by

$$n_s = n_{p0} \exp(q\psi_s/kT) \quad \text{and} \quad p_s = p_{p0} \exp(-q\psi_s/kT) \quad \text{(C-16)}$$

The potential distribution in the space-charge region is

$$\psi = \psi_s(1 - x/W) \tag{C-17}$$

where ψ_s is the surface potential (cf. surface field-effect in Section 4). The

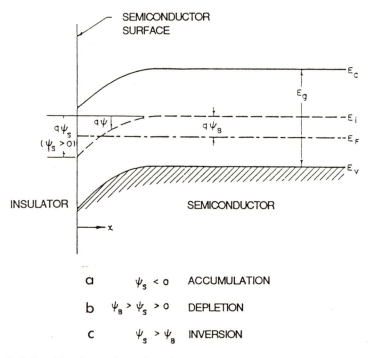

Figure C-4. Band bending at the surface of a p-type semiconductor. The zero potential is in the bulk of the semiconductor and is measured against the intrinsic Fermi level E_i. The surface potential ψ_s is shown as positive. (Reprinted from Ref. 4, p. 430 with permission of Wiley.)

direction of band bending depends on the polarity of ψ (Figure C-4). A special *flat-band* condition exists for $\psi = 0$ when no bending occurs. The analogy with a potential of zero charge in electrochemistry is clear.

References for Appendix C

1. S. M. Sze, *Physics of Semiconductor Devices*, 2nd ed., Wiley, New York, 1981.
2. R. M. Warner, Jr. and B. L. Grung, *Transistors*, Wiley, New York, 1983.

D

Equivalent Electrical Circuits[1,2]

Dynamic behavior of sensors can be modeled conveniently by equivalent electrical circuits. The aim of this approach is to obtain mathematical description of the system and also some insight into the physical meaning of individual components. The dangers in using equivalent circuits are a tendency to oversimplify and the fact that the same equivalent electrical circuit may fit several different models.

The simplest equivalent circuit models consist of capacitors and resistors, which represent specific *physical* elements in the model. Thus, for example, an electrode in solution can be modeled as a parallel combination of double-layer capacitance C_d, corresponding to the separation of charges at the interface, and a charge-transfer resistance R_{ct} corresponding to the kinetics of the electrolytic reaction. The electric current passes through an electrolyte solution of finite conductivity, so there is a solution resistor R_Ω in series. The final equivalent circuit of an electrode–solution interface then looks as in Figure D-1. If some additional information is known about the structure of such an interface (such as the separation of the diffuse and compact parts of the electrode double layer), additional circuit elements can be added. Thus, a more or less complex AC network is built.

Of course, many physical situations cannot be described by capacitors and resistors alone. As the circuit becomes more complicated the only possibility is to use an analog computer (either a true analog or digitally simulated) in which any nonlinear elements, function generators, and function multipliers/dividers can be used to describe the physical model. The danger of oversimplification and ambiguity in interpretation is even greater here.

Once the equivalent electrical circuit has been proposed, the next task is to determine its electrical response over the frequency range of interest.

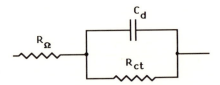

Figure D-1. Model equivalent circuit diagram representing double-layer capacitance C_d, charge-transfer resistance R_{ct}, and solution resistance R_Ω in an electrochemical experiment.

It is helpful to realize that the currents passing through the individual elements can be expressed by a generalized Ohm's law

$$I = E/Z \quad \text{or} \quad I = EY \tag{D-1}$$

where E and I are frequency-dependent voltage and current while Z and Y are impedance and admittance, respectively. The equivalent circuit shown in Figure D-1 is now examined. The simplest form of excitation signal is a small-amplitude (about 10 mV) sinusoidal voltage frequency which can be varied continuously. Physical intuition tells us that for any value of R_{ct} a high frequency can be found that will make the impedance of the double-layer capacitor negligibly small by comparison. On the other hand, for very low frequencies that impedance would be much higher than R_{ct}. Thus the effective overall impedance depends on frequency. Since the individual partial impedances become more or less prominent relative to each other, the frequency behavior conveys information about the structure of the equivalent circuit. It is sometimes advantageous to use other forms of excitation signal. Thus, a square-wave or triangular signal can simplify the interpretation of the characteristics of the transfer function. It is even possible to use white noise as the excitation signal[3] and carry out the interpretation in the frequency domain. The main advantage of this approach is speed.

The simplest and oldest technique (if not the most convenient) of experimentally evaluating the impedance is to use an impedance bridge or AC analyzer. These instruments yield values of impedance Z_b and phase

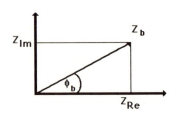

Figure D-2. Definition of the real (Z_{Re}), imaginary (Z_{Im}), and absolute (Z_b) impedance and phase angle ϕ_b.

angle ϕ_b, or of in-phase (R_b) and out-of-phase (C_b) components of the impedance (Figure D-2). The task is then to interpret the individual model elements in terms of these measured parameters. If the model consists of several impedance elements, different frequencies must be employed to make these individual impedances negligible (or dominating) with respect to each other. In order to do this it is necessary to convert a series capacitor/resistor combination into a parallel combination, and *vice versa*. A simple analytical procedure is outlined in the following section.[4]

D.1. Conversion of Equivalent Circuits

First we note that partial generalized impedances and admittances [equation (D-1)] add according to the same rules as ordinary (i.e., DC) resistances and conductances. Thus, two impedances in parallel add as admittances, etc. The next task is to establish the relationship between an equivalent parallel and series combination of R and C (Figure D-3). This can be done either graphically or analytically. The impedance of the series combination is a complex number

$$Z = R_s + 1/j\omega C_s \tag{D-2}$$

where ω is the angular frequency and $j = \sqrt{-1}$. For the parallel combination

$$1/Z = 1/R_p + j\omega C_p \tag{D-3}$$

We now define for series (s) and parallel (p) combinations

$$W_s = (\omega C_s R_s)^2 \quad \text{and} \quad W_p = (\omega C_p R_p)^2 \tag{D-4}$$

The admittance of the series combination is

$$1/Z = \frac{j\omega C_s}{j\omega C_s R_s + 1} \tag{D-5}$$

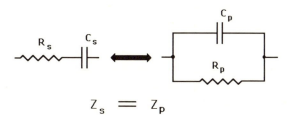

$$Z_s = Z_p$$

Figure D-3. Conversion of series to parallel *RC* combination. (Reprinted from Ref. 6, p. 630 with permission of Wiley.)

When both denominator and numerator are multiplied by $-(jW_s - 1)$, the real and imaginary components of equation (D-5) separate,

$$1/Z = \frac{\omega^2 C_s^2 R_s}{(\omega C_s R_s)^2 + 1} + \frac{j\omega C_s}{(\omega C_s R_s)^2 + 1} \tag{D-6}$$

or [with equation (D-4)]

$$1/Z = \frac{W_s/R_s}{W_s + 1} + \frac{j\omega C_s}{W_s + 1} \tag{D-7}$$

For the two combinations to be identical, the real and imaginary components of their impedances must be equal for all frequencies. Thus, for the real component, equations (D-3) and (D-7) yield

$$1/R_p = \frac{W_s/R_s}{W_s + 1} \quad \text{or} \quad R_p = R_s(W_s + 1)/W_s \tag{D-8}$$

and for the imaginary component

$$C_p = C_s/(W_s + 1) \tag{D-9}$$

A similar analysis can be performed for the series combination starting from the parallel combination. The resulting expressions are

$$R_s = R_p/(W_p + 1) \quad \text{and} \quad C_s = C_p(W_p + 1)/W_p \tag{D-10}$$

D.2. Complex Impedance Plane Analysis

The commonest way of presenting experimental data is in the form of a Cole–Cole plot (or Nyquist plot) in which the imaginary component is plotted against the real component with frequency as parameter. A Cole–Cole plot for the equivalent circuit given in Figure D-1 is shown in Figure D-4. In this case the impedance is

$$Z(\omega) = R_\Omega + Z_p(\omega) \tag{D-11}$$

The two frequency limits are now examined. At very high frequencies the admittance of the double-layer capacitor C_d shunts the charge-transfer resistor R_{ct}, so that the overall impedance is

$$\lim_{\omega \to \infty} Z(\omega) = R_{ct} \tag{D-12}$$

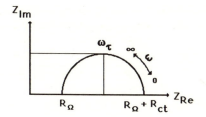

Figure D-4. A typical Cole–Cole plot corresponding to the equivalent circuit shown in Figure D-1.

At low frequencies the impedance of the double-layer capacitor is infinitely high, in which case

$$\lim_{\omega \to 0} Z(\omega) = R_{ct} + R_{\Omega} \qquad \text{(D-13)}$$

These are the two intercepts of the semicircle with the real axis in Figure D-4. Thus the semicircle describes the parallel combination of C_d and R_{ct},

$$Z_p(\omega) = \frac{R_{ct}}{j\omega R_{ct} C_d + 1} \qquad \text{(D-14)}$$

Separation into real and imaginary components is done similarly as for equation (D-6), multiplying both the numerator and denominator by $j\omega R_{ct} C_d - 1$,

$$Z_p(\omega) = \frac{R_{ct}}{(\omega R_{ct} C_d)^2 + 1} - j\frac{\omega R_{ct}^2 C_d}{(\omega R_{ct} C_d)^2 + 1} \qquad \text{(D-15)}$$

$$\underbrace{\hspace{2cm}}_{\text{real}} \qquad \underbrace{\hspace{2cm}}_{\text{imaginary}}$$

The frequency can be selected such that the real and imaginary components are equal. This point corresponds to the frequency at the top of the semicircle in Figure D-4,

$$\frac{R_{ct}}{(\omega_\tau R_{ct} C_d)^2 + 1} = \frac{\omega_\tau R_{ct}^2 C_d}{(\omega_\tau R_{ct} C_d)^2 + 1} \qquad \text{(D-16)}$$

from which

$$\omega_\tau = (RC)^{-1} \qquad \text{(D-17)}$$

or the time constant $\tau = RC$. Hence the time constant for a given impedance is identified with the top of the semicircle in the Cole–Cole plot. It can be shown that for N parallel combinations in series the Cole–Cole plot consists of N semicircles with their corresponding time constants. It is this visual aspect of these plots which makes them attractive for the interpretation of impedance data. However, often the individual semicircles are not well resolved and their assignment to physical impedances requires additional experimental data, such as concentration or temperature dependence.

References for Appendix D

1. A. J. Bard and L. R. Faulkner, *Electrochemical Methods, Fundamentals and Applications*, Wiley, New York, 1980.
2. D. D. Macdonald, *Transient Techniques in Electrochemistry*, Plenum Press, New York, 1977.
3. A. Bezegh and J. Janata, *J. Electrochem. Soc.* 133 (1986) 2087.
4. B. B. Damaskin, *The Principles of Current Methods for the Study of Electrochemical Reactions*, McGraw-Hill, New York, 1967.

E

Terminology

The science of chemical sensing is quite new and the terminology has not yet been firmly established. This appendix contains a list of terms which are used most often, together with the author's interpretation of their meaning.

Accuracy.   A characteristic of the measurement. If it is unbiased (i.e., the mean of several measurement repetitions approaches the true value) and well-reproducible, then the measurement is called accurate.

Actuator.   A device which uses the signal from the sensors to perform some *action.* An example is an alarm (with a smoke detector) or a hydraulic valve (coupled to a pH sensor).

Detection limit.   The concentration at which the mean value of the output signal is equal to two standard deviations.

Detector.   A device which indicates the presence (or absence) of the chemical species above (or below) a predetermined threshold value. There is no explicit quantitative relationship between the output and stimulant. An example is a smoke detector.

Determinand.   A species of interest.

Dosimeter.   A device which measures the dose. It provides a *time integral* of exposure to a given species. The chemical interaction is usually irreversible.

Dynamic range.   The range of concentrations in which the sensitivity is greater than zero.

Lifetime.   The usable period of the sensor. It must be specified whether "shelf" or "in-use."

Reproducibility.   An alternative expression for the dispersion of the signal, characteristic of the measurement, and usually expressed as variance, standard deviation, or coefficient of variation.

Selectivity.   The ability of the device to measure one chemical component in the presence of others.

Sensitivity. The slope of the response curve expressed as output per unit concentration. An example is ∂(frequency)/(ppm of gas) in piezoelectric sensors.

Sensor. A device for translating information about the concentration of a chemical species into a readily accessible signal (usually electrical). It responds to changes of that parameter in a bipolar way, i.e., the output signal follows changes in concentration, up and down, in line with some explicitly known functional relationship.

Stability. The percent of change of the baseline and/or sensitivity in time.

Time constant. The time at which the output reaches 63% ($1/e$) of its final value in response to a step change in concentration.

Transducer. A physical part of the sensor, detector, or dosimeter that amplifies the primary signal to a usable level. An example is a transistor.

Index